Advances in Human Factors/Ergonomics, 8

Human Aspects of Occupational Vibration

Advances in Human Factors/Ergonomics

Series Editor: Gavriel Salvendy, Purdue University, West Lafayette, IN 47907, U.S.A.

Vol. 1. Human–Computer Interaction (G. Salvendy, Editor)

Vol. 2. Human–Computer Dialogue Design (R.W. Ehrich and R.C. Williges, Editors)

Vol. 3. Expertise Transfer for Expert System Design (J.H. Boose)

Vol. 4. Engineering Physiology (K.H.E. Kroemer, H.J. Kroemer and K.E. Kroemer-Elbert)

Vol. 5. Human Factors Testing and Evaluation (D. Meister)

Vol. 6. Applications of Fuzzy Set Theory in Human Factors (W. Karwowski and A. Mital, Editors)

Vol. 7. Human Reliability (K.S. Park)

Vol. 8. Human Aspects of Occupational Vibration (D.E. Wasserman)

Advances in Human Factors/Ergonomics, 8

Human Aspects of Occupational Vibration

Donald E. Wasserman

Human Vibration Consultant and Director of Engineering, National Center for Rehabilitation Engineering, Wright State University, Dayton, OH 45435, U.S.A.

ELSEVIER
Amsterdam – Oxford – New York – Tokyo 1987

ELSEVIER SCIENCE PUBLISHERS B.V.
Sara Burgerhartstraat 25,
P.O. Box 211, 1000 AE Amsterdam, The Netherlands

Distributors for the United States and Canada:

ELSEVIER SCIENCE PUBLISHING COMPANY INC.
52, Vanderbilt Avenue
New York, NY 10017, U.S.A.

ISBN 0-444-42728-7 (Vol. 8)
ISBN 0-444-42396-6 (Series)

Printed in The Netherlands

To Helen, Melissa, and Sherri without
whose understanding, help and patience
this book could never have become a reality;
and to Dr. William Taylor, venerable friend, colleague,
and inspiration in his lifelong quest to rid the workplace of VWF.

Preface

Several years ago I entered the field of occupational vibration. During most of the time I had the unusual freedom, opportunity and challenge of setting the scientific course of virtually all aspects of hand-arm and whole-body occupational vibration studies (engineering, medical, epidemiological, industrial hygiene, field studies, and laboratory studies) while working for my former employer, the U.S. Government's National Institute for Occupational Safety and Health (NIOSH). From this effort emerged two distinct study programs; one in whole-body vibration, the other in hand-arm vibration. With the actual performance of several of these studies and their results published there emerged a long overdue sense of awareness and appreciation for the depth of occupational vibration problems in the U.S.; a sense of awareness which has long existed and attained venerability throughout Europe and Scandinavia.

The intent of this book is to try to bring together for the reader various aspects of occupational vibration; epidemiology, engineering, industrial hygiene, major medical and physiology studies, and human performance studies. The attempt is not to be exhaustive, since I fully recognize that any one of these areas would require a complete book unto itself, and several experts in each area, rather this text should be a catalyst for further investigation by the interested reader. Each chapter is virtually complete unto itself, supplemented with numerous references to aid the reader to go beyond the text.

By design my hope is that this is a practical book containing sufficient theory of a complex subject and yet intended for multiple audiences (e.g. human factors engineers, tool engineers, industrial hygienists, ergonomists, bioengineers, nurses, students, etc.) recognizing the differing backgrounds of each. This is not a medical text on the subject, although it does contain sufficient medical/epidemiological background for an understanding of the subject. The overall theme is engineering and industrial hygiene, namely to recognize, measure, evaluate, and control occupational vibration problems. For the most part, the text attempts to use simplifications of technical terms where possible, but in a few instances a knowledge of basic algebra and calculus is necessary. Also, this book by intent is concerned mainly with occupational hand-arm and whole-body vibration and thus the plethora of military studies concerning the latter are mentioned and referenced with a minimum of comment and within reach of the interested reader to pursue the topic more fully.

A text of this magnitude and complexity touching upon so much is never developed in a vacuum. I am indebted to several of my national and international colleagues for our numerous discussions and interactions much of which forms to a large extent the important issues discussed in this book, particularly: Mr. G. Rasussen (B&K Corp.), Dr. H.E. Von Gierke (USAFAMRL), Dr. J.C. Guignard (U.S.), Mr. J. Barton (Caterpillar Tractor Co.), Ms. V. Behrens (NIOSH epidemiologist), Dr. I-M. Lidstrom (Sweden), Dr. A.M. Seppalinen (Finland), Dr. D. Simic (Yugoslavia), Dr. L. Kazarian (USAFAMRL), Mr. W. Kelley (ACGIH), Mr. T. Doyle (NIOSH), Dr. T. Wilcox (NIOSH), Dr. H. Hursh (John Deere Corp.), Dr. P. Pelmear (Canada), Dr. S. Samueloff (Isreal), Dr. D. Reynolds (US), Dr. A.J. Brammer (Canada), Dr. M. Griffin, (U.K.), Dr. A. Irwin (U.K.), and my long time friend and colleague from Scotland Dr. W. Taylor. Finally, I am indebted to my current employer, pioneer physiologist and biomedical engineer Dr. Jerrold Petrofsky, Director of NCRE/Wright State University and to my editor Dr. G. Salvendy of Purdue University both of whom have encouraged me to write this book.

Donald Wasserman
National Center for Rehabilitation Engineering
Wright State University
Dayton, Ohio

CONTENTS

Chapter 1

Introduction to Occupational Vibration

1.1 Introduction to Occupational Vibration

Probably during the time when ancient man first assembled his shelter and used rocks as a hammer he noticed the sting of vibration impacts ringing through his unprotected hands. Similarly during the time of the ancient Phoenicians when man first took to the sea in his search of goods, the debilitating and often incapacitating effects of low frequency vibration were most likely known. Beginning with the industrial revolution, the worker became exposed to vibration in the occupational environment. Less than a century ago, disorders of the hands and fingers began appearing among men who worked with vibrating hand-held tools.

The advent of high performance aircraft marked the real beginning of studies on how vibration affects man's ability to perform work. It was soon apparent that the severe high-energy transient vibration associated with aircraft operation had a serious effect upon air crew performance. Studies of workers exposed to vehicular operation showed early onset of fatigue and potential impairment of the operator's ability to perform effectively in the workplace.

The potential problem of vibration is best exemplified by assuming that if a person works at the same vibration-susceptible job for 30 years, 50 weeks per year, at a 30-hour work week, then he can receive up to 45,000 hours of possible vibration exposure! In the United States alone some 8 million workers are estimated to be exposed to occupational vibration (1).

If the effects of noise is excluded from consideration, then the worker can be exposed to one or both of two types of mechanical vibration: (a) "whole body" vibration, or vibration transmitted to the entire body through some supporting structure, such as a vehicle seat, in a truck, bus, farm vehicle, or heavy equipment vehicle, for example; (b) "segmental" or "hand-arm" vibration, usually referring to vibration applied locally to specific body parts, such as the limbs, by vibrating chain saws, pneumatic or electrical tools, for example. An example of how workers can be exposed to both hand-arm and whole-body vibration is while using a road ripper or jack hammer. Some workers prefer to hold the jack hammer away from their bodies while working by extending their arms (segmental vibration); while other workers tend to lean against the hammer with their lower abdomen while still holding the hammer close to their body thereby coupling vibration energy to their body trunk (whole-body) as well as to their hands and arms.

1.2 Vibration in the Workplace

Of the estimated 8 million vibration exposed workers in the U.S. approximately 7 million are exposed to whole-body and 1 million are exposed to hand-arm vibration (1). Tables 1.1 and 1.2 give examples of a variety of occupations and the usual type of vibration found (2,3). However, the reader should not be led into thinking that these tables are all inclusive, for example, whole-body vibration is also found in large mold shakeout processes, fork lift trucks, and overhead cranes in foundries, yet hand-arm vibration is also found in the casting cleaning operations where workers use pneumatic chipping and grinding hand tools. The important thing to realize in most industrial operations is one must know the sequence of processes which produce the final outcome or product, and then to relate how vibration interacts with these processes. More often than not, in industrial plants, vibration will be found in specific processes and no vibration will be found in other processes. It is not enough to know just that vibration appears at a given plant site, one should know how and why it is used and why it is a part of the overall flow process. But just observing the overall flow process to produce a product in one plant, is no guarantee the situation will be the same in

TABLE 1.1

EUROPEAN INDUSTRIES IN WHICH CLINICAL EVIDENCE OF OVEREXPOSURE OF WORKERS TO VIBRATION HAS BEEN REPORTED*

Industry	Type of Vibration	Common Vibration Sources
Agriculture	Whole-body	Tractor operation
Boiler Making	Segmental	Pneumatic tools
Construction	Whole body	Heavy equipment vehicles
	Segmental	Pneumatic drills, jackhammers
Diamond cutting	Segmental	Vibrating hand tools
Forestry	Whole body	Tractor operation
	Segmental	Chain saws
Foundries	Segmental	Vibrating cleavers
Furniture manufacture	Segmental	Pneumatic chisels
Iron and steel	Segmental	Vibrating hand tools
Lumber	Segmental	Chain saws
Machine tools	Segmental	Vibrating hand tools
Mining	Whole body	Vehicle operation
	Segmental	Rock drills
Riveting	Segmental	Hand tools
Rubber	Segmental	Pneumatic stripping tools
Sheet metal	Segmental	Stamping equipment
Shipyards	Segmental	Pneumatic hand tools
Stone dressing	Segmental	Pneumatic hand tools
Textile	Segmental	Sewing machines, looms
Transportation (operators and passengers)	Whole body	Vehicle operation

*From reference 2.

TABLE 1.2

POTENTIAL OCCUPATIONAL VIBRATION EXPOSURES IN U.S. INDUSTRIES

Industry/Job Type	Type of Vibration	Common Vibration Sources
Truck driving	Whole body	Truck engine
Bus driving	Whole body	Bus engine
Heavy equipment operation	Whole body	Scrapers, loaders, bulldozers
Farm vehicle and tractor operation	Whole body	Farm tractors, harvestors
Foundries	Whole body	Mold shakeout, fork lift, trucks, overhead cranes
Fork lift operation	Whole body	Engine, fork lift system, crane and cab motion
Overhead crane operation	Whole body	Crane and cab motion
Textile machine operations	Whole body	Weaving machine
Metal: Refining	Whole body	Electric furnaces
Mills	Whole body	Rolling operation, fork lift trucks, overhead crane
Manufacturing	Whole body	Stamping operation
Machine tool operation	Whole body	Machine tool
Quarry worker	Whole body	Heavy equipment and machine
Mining (strip and underground)	Whole body	Heavy equipment and automatic mining machine
Vehicular body	Whole body	Stamping operation
Lumber (mills, sawplants, wood products)	Whole body	Stationary saws, cutter, sanders, etc.
Printing and publishing	Whole body	Press motion
Shoe manufacturing	Whole body	Stationary stitching machine
Lumber (tree cutting and debranching)	Segmental	Chain saws
Mining	Segmental	Jack leg-type drills, pneumatic picks and jack hammers
Construction	Segmental	Jack hammers, staplers, drills, saws, impact hammers, etc.
Foundries	Segmental	Chipping, grinding, scraping
Shipbuilding	Segmental	Grinding, caulking
Metal working	Segmental	Drills, sanders, grinders

*From reference 3

another plant producing the same product. In other words, all foundries are not necessarily the same, neither are all steel mills, etc. yet their products may look the same when completed.

In summary, there are many thousands of workers exposed to occupational vibration. What are these exposures? What are the potential medical or safety consequences of such exposure? What can be done to reduce or eliminate vibration impinging on the worker? Is all vibration necessarily bad or can it be used in a useful manner? The following chapters will attempt to address these and other questions.

REFERENCES*

1 D.E. Wasserman, D. Badger, T.E. Doyle and L. Margolies, Industrial Vibration - An Overview, J. Amer. Soc. Safety Engineers, 19 (1974) 38-43.

2 R.D. Soule, Chapter 26 - Industrial Environment - Its Evaluation and Control, NIOSH Syllabus, U.S. Gov't Printing Office, Washington, D.C. 1973.

3 D.E. Wasserman, Vibration - Occupational Diseases: A Guide to Their Recognition DHEW/NIOSH Public. 77-181, U.S. Gov't Printing Office, Washington, D.C. 1977.

Supplementary Bibliography

D.E. Wasserman and D.E. Badger, Vibration and the Worker's Health and Safety, DHEW/NIOSH Technical Report 77, Cincinnati, Ohio 1973.

J.C. Guignard and E. Guignard, Human Response to Vibration, Institute of Sound and Vibration Research, Report 373, Southampton, England, 1970.

*Footnote: Each chapter contains both cited (numbered) references and additional supplementary references to aid the reader. In succeeding chapters some of these references may be repeated as necessary.

Chapter 2

The Human Aspects of Occupational Vibration

2.1 INTRODUCTION

In this chapter the major effects of vibration on humans, particularly workers, are discussed in terms of epidemiology, occupational medicine, and performance measures. Since the characteristics of the vibration impinging on workers and the resulting health and potential safety effects are different for hand-arm (segmental) vibration versus whole-body vibration, each will be discussed separately.

2.2 OCCUPATIONAL HAND-ARM (SEGMENTAL) VIBRATION-MEDICAL EFFECTS (1-4)

2.2.1 History

In 1862 a French Physician named Dr. Maurice Raynaud wrote a now famous M.D. thesis entitled "Local Asphyxia and Symmetrical Gangrene of the Extremities" (5).

In this dissertation is found the first description of "a condition, a local syncope, where persons, who are females, see under the least stimulus one or more fingers becoming white and cold all at once. The determining cause is often the impression of cold. The cutaneous sensibility also becomes blunted and then anihilated." This is a condition of the fingers called "white finger" or "dead hand." The afflicted are about 5% - 8% of the general population (90% of these are females). Today this condition is now referred to as PRIMARY RAYNAUD'S DISEASE. The patient experiences blanching attacks which usually affect the fingers symmetrically and are relatively minor in the early stages of the disease, no so in the later stages. A small proportion (1% to 3%) may become progressively more severe and over a period of years the condition leads to blue cold fingers, where the skin starts to atrophy; followed by ulceration; and finally the fingers become gangrenous. Dr. Raynaud also observed an emotional or psychosmatic factor present in this "white finger syndrome"; the number and severity of blanching attacks increased during emotional stress. In 1911 Loriga (6) in Italy first described "vascular spasm" or white finger in the hands of pneumatic tool miners.

It was not until 1918 that we obtained the first comprehensive study and linkages of vibration to white finger syndrome, described by pioneer occupational physician Dr. Alice Hamilton. Well known for her toxicology and other pioneering occupational studies (7,8). As a U.S. Government physician, Dr. Hamilton went to the olitic limestone area of Bedford, Indiana in the U.S. where she examined 28 stonecutters/carvers who use vibrating pneumatic tools (9); there she found 89% had RAYNAUD'S PHENOMENON OF OCCUPATIONAL ORIGIN. This study was and still remains one of the classic studies of the day. Without the benefit of modern medicine, she correctly concluded that it was a combination of vibration, cold temperatures, and ergonomics (e.g., the manner in which the tools were held and used) which were responsible for the malady. (For the interested reader, the Appendix contains exerpts from the original 1918 Hamilton Study.)

There have been reports of Raynaud type symptomology before and after World War II. With the advent of the gasoline powered chain saw in the early 1950's and 60's, the "white finger" condition was recognized as arising from the effects of prolonged exposure to vibration entering the hands (1,3) originally referred to as Raynaud's Phenomenon of Occupational Origin the condition was now called VIBRATION INDUCED WHITE FINGER or VWF. However, with further research into VWF, signs and symptoms associated with vibrating tools were reported in other systems such as nerves, bones, joints, and muscles. Thus, the complex of VWF and these associated disabilities has become known as VIBRATION SYNDROME (1-3).

In the literature the terms VWF, Vibration Syndrome, and another term Traumatic Vasospastic Disease (TVD) are currently used nearly interchangeably.

2.2.2 VWF/vibration syndrome symptomology and classification

Some medical conditions may result in symptoms similar to vibration syndrome making detection and the physician's differential diagnosis very difficult. Table 2.1 summarizes these other conditions (2,10).

Taylor and Pelmear (2) have described and stratified and staged the clinical manifestations of vibration syndrome (shown in Table 2.2). (There are other classification schemes, but we chose to describe this well known system.) The first symptoms of the malady are intermittent tingling and/or numbness of the fingers, without interfere with work or other activities. Later, the workers may experience attacks of finger blanching usually confined at first to a single fingertip. With additional vibration exposure, attacks may extend to the base of the finger. Cold often triggers attacks but there are other factors involved, such as central body (core) temperature, metabolic rate, vascular tone of the vessels (especially in the early morning), and emotional state. Attacks usually last 15 to 60 minutes. In advanced stages, attacks may last as long as 2 hours. Recovery from attacks can also be painful, starting with a red flush, usually seen in the palm, advancing from the wrist towards the fingers.

Based on a clinical observation job history and medical interview, the physician places the worker into one of the Taylor/Pelmear classification categories shown in Table 2.2. Generally, stage 1 and stage 2 attacks occur mainly in the winter and especially during the early morning, either at home or when driving to work, especially when the hands contact cold steering

TABLE 2.1

Exclusion Criteria and Differential Diagnosis for Raynaud's Phenomenon. (from Reference 14)

Primary Raynaud's disease	Constitutional white finger
Secondary (Raynaud's Phenomenon) Connective-tissue disease	Scleroderma, systemic lupus erythematous, rheumatoid arthritis, dermatomyositis, polyarteritis nodosa, mixed connective-tissue disease
Trauma Direct to extremities	Following injury, fracture, or operation; of occupational origin vibration; frostbite and immersion syndrome
To proximal vessels by compression	Thoracic outlet syndrome (cervical rib, scalenus anterior muscle), costoclavicular and hyperabduction syndromes
Occlusive vascular disease	Thromboangiitis obliterans, arteriosclerosis, embolism, thrombosis
Dysglobulinemia	Cold hemagglutination syndrome: cryoglobulinemia, macroglobulinemia
Intoxication	Acro-osteolysis, ergot, nicotine
Neurogenic	Poliomyelitis, syringomyelia, hemiplegia

wheels. People working outside in cold weather, such as forestry workers, are most prone to early morning attacks. Studies have shown that as the exposure time increases, the number of attacks tends to increase (1,4). During stage 2, workers may report interference with or limitation of activities outside their work (e.g., hobby activities such as gardening, fishing, swimming, outdoor sports, woodworking, etc.). These activities have one common factor: in the cold, they are more likely to trigger an attack.

TABLE 2.2

Stages of Vibration White Fingers.

(Taylor-Pelmear System, Reference 2)

Stage	Condition of Fingers	Work and Social Interference
00	No tingling, numbness, or blanching of fingers	No complaints
OT	Intermittent tingling	No interference with activities
ON	Intermittent numbness	No interference with activities
TN	Intermittent tingling and numbness	No interference with activities
1	Blanching of a fingertip with or without tingling and/or numbness	No interference with activities
2	Blanching of one or more fingers beyond tips, usually during winter	Possible interference with activities outside work; no interference at work
3	Extensive blanching of fingers; frequent episodes in both summer and winter	Definite interference at work, at home, and with social activities; restriction of hobbies
4	Extensive blanching of most fingers; frequent episodes in both summer and winter	Occupation usually changed because of severity of signs and symptoms

In stage 3, the attacks occur both in summer and winter. There is interference with work, particularly if it takes place outdoors, such as forestry and construction. But also indoors as well, such as difficulty with fine work, in picking up small objects. Patients also experience difficulty in buttoning and zippering clothing; inability to distinguish between hot and cold objects; and clumsiness of fingers with increasing stiffness of the finger joints and loss of manipulative skills.

Stage 4 is the most severe of all. Not only is there interference with work, social activities, and hobbies but usually workers must change their occupation. Medically in this stage there are advanced changes in the arteries of the fingers, leading in many cases to complete obstruction of the arteries and in a very few cases these progress to gangrene.

For the most part vibration syndrome in these later stages is considered by most to be irreversible (1-4). The increasing stages appears to arise from the cumulative effect of vibration transmitted to the hands arising from the regular and prolonged use of vibrating tools found in the workplace.

Epidemiologists describe the length of the initial symptom-free period of vibration exposure (i.e., from first vibration exposure to the first appearance of a white finger) as the LATENT PERIOD or INTERVAL. It is related to the intensity of the vibration--the shorter the latent period; the more severe the resulting vibration syndrome is, if vibration exposure continues.

2.2.3 Epidemiology

Traditionally hand-arm vibration epidemiology/measurements studies have concentrated on three tool categories: a) gasoline powered chain saws (1-4, 11-14), (because of the unbalances in the engine considerable vibration is present); b) some electrically operated tools (2,3,15) (e.g., drills, impact hammers, grinders, sanders, pedestal grinders, etc.); c) pneumatically operated tools (1-3,6,9,10,16-21) (e.g., chipping tools, grinders, jack-hammers, riveters, jack-leg type mining drills, etc.).

Typical studies of conventional chain saw users have shown VWF prevalences of 6-89% (2); (after medical exclusions) with a range of latent periods of 1-5 years. Recent chipping and grinding tool studies in foundries and shipyards (10,23-25) have shown prevalences of VWF (after medical exclusions) ranging from 19-47% with a range of latent periods of 2-17 years (in a NIOSH study of 385 vibration exposed workers). Typical acceleration levels of conventional chain saws (without antivibration A/V protection) can range as high as 30 g_{rms}, in a frequency range of approximately 10-2000 Hz (1-3). Similarly, pneumatic chipping tool vibration levels can exceed 2,000 g_{rms} depending on the tool type, use, condition, etc. (4,10,26,27) (see Tables 2.3 and 2.4).

The reader will note that despite the initial historic studies of Alice Hamilton, much of the VWF epidemiological work prior to the 1970's was not performed in the U.S. Ironically it was a study in the U.S. by Pecora et al (28) which stated that VWF "may have become an uncommon occupational disease approaching extinction in this country (U.S.)," after surveying some 18 factories, interviewing physicians, workers, etc. However, a later study by NIOSH discovered a common reticence on the part of workers to report VWF symptoms because of potential loss of their employment (29). Other aforementioned NIOSH studies (10,23,24) have conclusively demonstrated that VWF does indeed exist in the U.S. Even more ironic is that the NIOSH team repeated the Hamilton study, 60 years later, at the same location in Bedford, Indiana and found virtually the same VWF prevalence (80%) and that neither the work conditions nor the chipping tools had changed; in fact the measured vibration levels on the Bedford limestone hammers ranged from 494 g_{rms} on the chisels to 205 g_{rms} on the tool barrels, where hands grip the tools (30).

TABLE 2.3

Summary of Acceleration Levels for Grinders.

(from Reference 10)

Wheel Type	Handle	Total Equivalent Acceleration G(rms)	Additional Description
		Horizontal Grinder	
Course radial	Right	0.74	Cast iron, grinder condition excellent
	Left	1.61	
Fine radial	Right	0.64	Cast iron, grinder condition excellent
	Left	0.79	
Flared cup	Right	1.71	Cast iron, grinder condition excellent
	Left	2.09	
Cup	Right	27.83	Ductile, grinder condition average
	Left	39.75	
	Right	26.42	Gray iron, grinder condition average
	Left	44.56	
Cone	Left	60.04	Ductile, grinder condition average
	left	79.07	Gray iron, grinder condition average
		Vertical Grinder	
Sanding	Right	0.97	Cast iron, grinder condition excellent
	Left	0.56	
New	Right	35.79	High-alloy gray iron, grinder condition average
	Right	49.82	Gray iron, grinder condition average
Old	Right	28.74	High-alloy gray iron, grinder condition average
	Left	22.63	
	Right	49.75	Gray iron, grinder condition average
	Left	24.80	
Cutting	Right	21.37	High-alloy gray iron, grinder condition average
	Left	22.45	
	Right	15.56	Ductile, grinder condition average
	Left	26.98	

TABLE 2.4
Summary of Acceleration Levels for Chipping Hammers.
(from Reference 10)

Work Site	Chipping Tool	Accelerometer Position	Throttle	Type of Acceleration G(rms)	Material Chipped	Chipping Operation
Foundries	Hammer A	Chisel Handle	Full	2,388 30.5	Cast iron	Slot chipping
	Hammer B	Handle	Full	35.8	Cast iron	Slot chipping
	Hammer C	Handle	Full	12.3	Cast iron	Slot chipping
Shipyard	Hammer D	Handle	Full	66.0	Cast iron	Slot chipping
		Chisel	1/2-3/4	194	Nickel-Aluminum-Bronze	Continuous
		Handle		3.8		
		Chisel	1/2-3/4	203	Mild B steel	Continuous
		Handle		4.1		

2.2.4 Major problem areas related to VWF

At this writing there are three major problem areas related to VWF: a) little is known about its etiology and physiological basis; b) there are no reliable objective medical screening tests for VWF; c) there are no long term medical cures for VWF; currently used medical modalities are pallative at best.

We will try to briefly examine these problem areas, recognizing that this presentation is not meant to be a medical treatise on this subject.

2.2.4.1 Etiology and physiological basis of VWF

Unfortunately little is known about the etiology and physiological basis of VWF or how such parameters as vibration acceleration, velocity, resonances, frequency spectrum, or mechanical coupling alter the acquisition and/or the course of VWF. The best thinking to date asserts that VWF results most likely from the cumulative effect of vibration-induced microtrauma to the nerves and blood vessels of the hands due to regular continued use of vibrating hand tools (1-4,10,23). Since VWF has two major components associated with it, namely, neurological and peripheral vascular the question remains as to what is actually being affected or destroyed, is it the nerves and/or the blood vessels? No one knows whether the primary cause of the vasospasms is neurologic or peripheral vascular. There may be direct damage to the sensory nerves or it may be that peripheral vascular impairment results in the ischemia, or finger blanching. Behrens et al (10) describes the probable cause of finger blanching succinctly:

"The development of finger blanching in workers exposed to vibration may be influenced by some combination of the following factors: (A) increased sympathetic vasomotor tone secondary to stimulation of the sympathetic system by vibration and possibly by noise and cold (31); (B) local effects on the vessel, such as hypertrophy of the medial muscular layers of the arterioles (32), or vibration-induced hyper-responsiveness of the smooth muscle of the vessels to the vasoconstrictive effects of circulating noradrenaline (33); (C) adverse effects from vibration on the arteriovenous anastomosis (34) and (D) intimal damage to the vessel due to vibration-induced shear stress (35).

In support of factor A, exposures to vibration, noise, and cold have all been demonstrated to induce constriction of the arteries of the hand (31). In support of factors B and C, biopsy specimens from the fingers of vibration-exposed workers with vibration syndrome have shown intimal fibrosis (32,36) as well as alterations of the arteriovenous anastomoses (34) in the skin.

Recent articles have reported elevated blood viscosity in workers suffering from vibration syndrome (37) as well as in other forms of Raynaud's phenomenon (38). The increased blood viscosity may be due to substances produced secondary to vascular endothelial damage, as other conditions associated with vascular damage are also accompanied by increased blood viscosity. The increased blood viscosity may not cause the vessel spasm, but it may further decrease blood flow to the digits during attacks of vasospasm" (10).

2.2.4.2 Screening tests for VWF

Many tests for VWF have been tried through the years and unfortunately these tests have all failed to adequately screen for VWF. Their major problem appears to stem from three things: a) some of the tests will distinguish between vibration and control (nonvibration) populations, but these same tests fail to screen for VWF on an individual basis. b) Since the etiology of the disease process is not well understood, it is very difficult to design medical screening tests with sufficient sensitivity to prescreen worker's who might get VWF as a result of working with vibrating hand tools; c) it is not possible to provoke a VWF attack at will, thus tests specifically designed to

TABLE 2.5

VWF Objective Test*.

PERIPHERAL VASCULAR

- Skin temperature
- Infrared thermography
- Brachial artery compression tests
- Cold-pressor recovery index
- Allen's test
- Plethysmography
- Finger plethysmography + provocative cooling
- Nailbed test (Lewis-Prusik)
- Whole-blood viscosity index
- Iontophoresis
- Brachial arteriography

NEUROLOGICAL

- Two point discrimination
- Light touch
- Pain threshold
- Temperature changes
- Depth sense
- Vibration perception thresholds
- Vibration temporary threshold shifts
- Electromyography
- Galvanic skin response
- Conduction velocity

CHANGES IN THE MUSCULOSKELETAL SYSTEM AND MOTOR DYSFUNCTION

- Grip force
- Pinching power
- Tapping ability
- Tonic vibration reflex
- Excretion of urinary hydroxyproline

X-RAYS OF ELBOW, WRIST, AND PHALANGES

- Roentgenograms of bones in elbow, wrist, and fingers

CNS INVOLVEMENT

- Excretion of urinary catecholamines

OTHER TESTS

- Finger circumference measurements

---

*Adapted from reference (10)

measure differences between normal vs abnormal fingers and hand conditions cannot be easily designed and used in the absence of the malady.

In general these test can be grouped into the categories given in table 2.5 (3,10).

Peripheral vascular tests: Since some of the signs and symptoms of VWF are finger blanching, it would seem reasonable that tests related to blood flow in the fingers and hands could be used as objective tests of VWF. Similarly, since skin temperature in the hands and fingers depend on peripheral blood flow this too would seem reasonable as an approach to an objective test. For example, in a study by Pelmear et al (2) of U.K. chain sawyers, the researchers compared skin temperatures using an infra-red camera and film to standard thermocouples after cooling of the hands at 7-8°C for 4 minutes and hands withdrawn for 5 minutes. The results were not encouraging, the technique did not distinguish controls from vibrated subjects nor could it be used to stratify VWF severity.

It would also seem reasonable that the rewarming time necessary to get VWF patients finger blood flow back to "normal" would be considerably longer than in control patients who don't have VWF. In one NIOSH study we developed a special photocell plethysmography instrument (39) which would detect blood flow in five fingers of each hand simultaneously. We evaluated the test under laboratory conditions using six Raynaud patients (40). The lab results were encouraging because it would appear we could distinguish between degrees of Raynaud severity in these laboratory patients. We then used this test in an extensive field study of 154 chipper and grinder workers and controls. Unfortunately, the results were inconclusive, but we were able to demonstrate a weak association between VWF, prolonged recovery time and blood flow (10). On-the-other hand Nielsen reported in a lab study that provocative finger cooling together with whole-body cooling of 18 females with primary Raynaud's disease had a significantly greater finger systolic blood pressure as compared to 22 normal females (41); he proposes this method as an objective test.

Neurological tests: These tests include measurements of two point discrimination, depth sense, light touch, pain threshold, temperature changes, depth sense, etc. These tests too would seem reasonable since the early stages of the malady include tingling, numbness and loss of touch in the fingers.

Two point discrimination was originally developed by Renfrew (42) and later used by Pelmear (1), and later by Carlson et al (10,43). The measurement consist of an instrument containing two sharp edges forming a V shape. Each finger is run over the V beginning at the intersection of V edges moving outward. The point at which the subject "feels" the separation of the V edges is noted. One could postulate that the further out along the V the subject indicates feeling the edges separate, the more the neurological damage and consequently the higher the VWF stage. A depth sense device was also developed by Carlson et al (43). Once again these devices appeared promising in the laboratory, but in the field these tests proved insensitive to stages below level 2 and false positives and negatives appeared in the results (10).

The use of cotton swatches brushed against various fingers and hand areas and using a sharp needle touching various parts of the fingers and palms together with temperature changes in the hands and fingers (e.g., having the patient determine which test surfaces are hotter or colder than other surfaces) were all used by Hamilton et al in the original Raynaud studies (9), however these too may have observer bias and uncontrolled variables (4).

Musculoskeletal tests: In an effort to quantitate diminished tactility sensitivity (as a possible objective test), Streeter (44) used vibration sensitivity from a hand dynamometer at various vibration frequencies of 30, 60, 120, 240, 480 Hz at three acceleration levels using various grip strengths (6, 12, 18 lbs. force). He found that sensitivity was frequency dependent and most pronounced at 60, 120, 240 Hz with grip strength insignificant and the higher accelerations producing decreased sensitivity loss. Changes in nerve conduction velocity have also been used by Seppalainen with moderate success due to the lack of specificity of test (45).

X-rays: The use of finger, hand, wrist, elbow x-rays as a potential diagnostic tool is based on two issues: a) sparse evidence that segmental vibration produces osteoporosis (e.g., the loss of calcium in the bones, often called "bone thinning") in the upper limb bones (1-4), and b) small cysts are formed on the finger and hand bones where the vibration energy from vibrating tools is highest (2). Unfortunately the NIOSH hand x-ray study of 205 vibration exposed and 63 control workers could not confirm these claims (10).

In summary, there have been many efforts to determine an objective test and/or a combination of objective tests for VWF. None to-date have been clinically adequate. Thus the research in this area and the quest for clinical objective tests continues (4,46). Today, the physician uses the workers medical history, work history and a concentration of tests and experience to identify and stratify VWF in the workplace.

2.2.4.3 Medical cures for VWF (47)

To date no long term cures for VWF exist, medical modalities work on the short term and are pallative at best. Most stage 2 and 3 patients are given vasodilators when reporting to their physicians. The results of attempting to dilate the peripheral vessels by direct action through chemotherapy has been unsuccessful and can be dangerous in some subjects because of its side-effects. The failure of chemotherapy seems to be due to the absence of a therapeutic agent which acts specifically on the digital arteries. Vasodilatation of the digital vessels by surgical sympathectomy results in immediate improvement in blood flow but lasts only a few months. The vasoconstrictive attacks return again with the sympathetic nerve supply still blocked or severed. There are claims of some success with plasmaphoresis (48). The most recent development is the use of prostaglandin which has a strong inhibitory effect on platelet aggregation and the capacity to dissolve recently formed platelet aggregation as well as prozosin and various calcium channel blockers (49,50). Some success has also been reported using biofeedback techniques (49-52).

2.2.4.4 VWF and carpal tunnel syndrome (53)

Medically, carpal tunnel syndrome, also called entrapment neuropathy, may be due to repeated mechanical trauma, edema, fibrosis, tuberculosis and other diseases, including rheumatoid arthritis. It is caused by pressure on the median nerve as it passes through the space formed by the bones of the wrist and the carpal ligament. Any space-occupying process may cause compression and malfunction of the median nerve or other nerves similarly confined.

Patients with carpal tunnel syndrome often complain of burning pain on the palm and the first three fingers of the hand. This pain may also radiate up the arm. The pain usually is more severe at night when numbness may develop. Electromyography (EMG) and nerve conduction tests across the wrist are used in the diagnosis.

Surgery of the carpal tunnel releasing the transverse part of the carpal ligament is the most common treatment when symptoms and signs persist. This condition can become incurable at some point during the progression of the malady (53,54). Vibrating hand tools can also produce carpal tunnel syndrome with or without VWF (55,58). A few researchers believe that the tool handle size has much to do with the occurrence of both VWF and carpal tunnel syndrome, namely, when workers use a large handle tool the vibration energy is spread over a much greater palm-finger surface than when workers use a small handle tool which when held in a pinch grip concentrates most of the vibration energy into the carpal tunnel region of the palm (59). This theory indeed may have some validity since, for example, in the previously cited NIOSH studies (10) virtually no carpal tunnel syndrome was found, yet there were very high prevalences of VWF and the tools had generally large handles. An important part of the physicians differential diagnosis thus is to separate what the worker may be afflicted with VWF or carpal tunnel syndrome or both, this is not trivial, it is very difficult to do.

2.3 Whole-body vibration

Unlike hand-arm vibration which appears to be inextricably linked to a definable biological end-point (e.g., VWF) whole-body vibration is not a specific stresser and appears to be a generalized stresser impinging on virtually the entire body. Unfortunately little is actually known about the chronic medical and epidemiological aspects of whole-body vibration (from 1-80 Hz). There is much more known about the subjective behavioral and performance aspects of subjects under whole-body exposure; however, the reader is to be cautioned that much of the subjective behavioral/performance studies to date have been done under controlled, laboratory conditions, trying to simulate military environments such as the cockpit of a jet aircraft or aboard navy ships simulating shipboard motion, using young physically fit military subjects which may or may not represent a typical worker (60). Also, the majority of these laboratory performance studies were performed using discrete sinusoidal vibration which may or may not represent the spectral conditions found in most workplace situations (61). The watch word is simply this, laboratory study conditions using either human or animal subjects and the results therefrom does not necessarily attempt to or represent the real world (e.g., workplace, truck cabin, etc.). These data should be used very cautiously with respect to real occupational situations.

2.3.1 Epidemiological and medical effects of whole-body vibration

Prompted by worker and union complaints NIOSH performed four morbidity (health) record studies on various U.S. worker populations: intercity bus drivers (62); long distance truck drivers (63); and two studies of heavy equipment operators (64,65). The bus drivers study showed a significance excess of venous, bowel, respiratory, muscular, and back disorders in a population of 1,448 interstate bus drivers as compared to two control groups: office workers and general population. This study concluded that the combined effects of body posture, postural fatigue, dietary habits, and whole-body vibration would appear to contribute to the occurrence of these disorders. The long distance truck driver study looked at 3,205 drivers and a control group of air traffic controllers who experience stress but no vibration. The study showed that the combined effects of forced body posture, cargo handling, improper eating habits and whole-body vibration were factors contributing to significant excess of vertebragenic pain, spinal deformities, sprains, strains and hemorrhoid disorders among truck delivers. The first heavy equipment operator study of several hundred health claims (64) showed that exposed workers had an elevated risk of male genital diseases, certain musculoskeletal diseases, ischemic heart disease, and obesity of non-endocrine origin. The study results led to the hypothesis that workers with musculoskeletal, heart, and obesity disorders might be leaving their jobs due to vibration exposure. A second follow-up study (65) could not confirm this hypothesis and actually showed that exposed and control workers both left their jobs when they had these disorders. In another study of 371 farm tractor operators, it was concluded that the low back pain, gastrointestinal problems and spinal difficulties caused by premature deformations of the verbetrae observed in exposed workers (65) were exacerbated by other factors, such as poor seating, poor posture, long irregular hours, etc. Epidemiology studies of helicopter pilots have been performed and low back pain has consistently been reported but not necessarily attributable to whole-body vibration alone but in combination with other factors such as poor posture due to poor seat design and orientation and long hours of helicopter vibration exposure (67-70).

Kelsey in the U.S. performed a series of epidemiology studies on car and truck driving (71,72) and concluded that those persons involved in driving, particularly truck drivers, were nearly three times more likely to develop acute herniated lumbar discs when compared to control subjects who never drove a truck.

Guignard and King (72) have chosen to classify the nebulous whole-body physiological effects into various systems, this ostensibly because of the lack of hard medical data between vibration exposure and biological response. This systems approach will be followed here, too.

a) Muscular activity and maintenance of posture: In the 1-30 Hz range human subjects have difficulty in maintaining posture and have increased postural swing and there is also a tendency to have a depressed jerk reflex in a wide frequency in up to 10-200 Hz (73).

b) Cardiovascular system effects: At frequencies below 20 Hz, there appears to be an increase of heart rate (74,75) during vibration exposure. A peripheral vasoconstrictor response seems to result in both human and in unsedated animals at moderate vibration levels (76), the latter phenomenon is not unlike locally applied vibration to the hands.

c) Cardiopulmonary effects: It would appear that whole-body vibration sets off the alarm in humans with increase in oxygen uptake, pulmonary ventilation and respiratory rate (73,77) and hyperventilation, particularly in the 0.5 g range, 1-10 Hz (78).

d) Metabolic and endocrinological effects: As a general reaction to stress changes in blood and urine biochemistry have been observed in both human and animal studies (79-81).

e) <u>Central nervous system effects</u>: Findings concerning CNS response have been mixed and varied (81). In general East European researchers claim vibration causes a general debilitating and malaise effect which is called "vibration sickness." A Polish study of agricultural and forestry workers describes this condition (82):

> "The first stage (of vibration sickness) is marked by epigastralgia, distension, nausea, loss of weight, drop in visual acuity, insomnia, disorders of the labyrinth, colonic cramps, etc. The second stage is marked by more intense pain concentrated in the muscular and osteoarticular systems. Objective examinations of the workers disclosed muscular atrophy and trophic skin lesions. It is apparent that it is difficult to determine the critical moment at which pathological changes set in, especially due to differences in individual sensitivity to vibration."

f) <u>Gastrointestinal system effects</u>: A series of laboratory studies using primates was performed as a joint effort between NIOSH and the U.S. Air Force AMRL vibration laboratory. In these studies primates were subjected to 12 Hz 1.5 g peak vertical whole-body vibration for 5 hours/day, 5 days/week for 130 hours. The results of necropsies indicated a general stress response, namely, extensive gastrointestinal bleeding and lesions for 10 exposed animals and none in the 13 control animals (83,84). Similar results have been seen by other researchers (85).

g) <u>Motion sickness effects</u>: Motion sickness (kinetosis) can occur at sea, in airplanes or space vehicles, or in land vehicles. It results in nausea, vomiting, general malaise and occurs in the 0.1-1 Hz frequency band with the patient recovering after the removal of the vibration stimulus and/or medication (81,86-90). Motion sickness appears to be worse at about 0.3 Hz (73,91), whereas at higher frequencies (above 1 Hz) even with high vibration levels, motion sickness does not result. It would appear that the centers associated with orientation and posture together with vestibule-cortical apparatus and head movement are most effected, whereas emotional state, when the last meal was consumed, etc. seem not play a leading role in the malady (91). It would also appear that when such very low frequencies are found during heavy equipment operation, motion sickness does not occur, presumably because the operator sees fixed land points; the low frequencies are not sustained and are transient in occurrence; and the operator is in control of the vibrating situation (92).

2.3.2 <u>The performance effects of whole-body vibration (73,93)</u>

There are a plethora of human studies during vibration exposure. The results of many when compared often show confusing and contradictory results leaving the reader confused. There are some common denominators, however, which may serve to clarify this matter. First, nearly all of the human performance studies thus far performed are short term exposures, usually using physically fit, young, male, (military) personnel as subjects. Many studies use a series at individual single frequency vibration in the vertical (z) direction at low to moderate (less than 0.5 g_{rms}) acceleration levels (because of possibly inflicting injury on the subjects). Unfortunately, with few exceptions, these studies do not necessarily resemble the workplace situation which contain multiple vibration frequencies and accelerations appearing simultaneously (e.g., from vehicle operation, varying road conditions, etc.); whose drivers are not necessarily young physically fit personnel in optimum condition; and whose work environment contains a combination of vibration, noise, heat, dust, etc. all simultaneously impinging on the worker as he works 8 hrs per day, year after year. The realities of this isolated laboratory research is that these studies represent true efforts attempting to isolate and determine the performance effects of whole-body vibration. The fear is that the results of these studies will be used to predict the effects of workplace exposures, for which most were never intended in the first place. Nonetheless, it is worthwhile to cautiously attempt to summarize some of the major studies and what they seem to indicate.

First, many studies are subjective with "operator comfort" as the subjective issue. Secondly, other studies are concerned with "performance effects" as the subjective issue. The latter portending the possible safety effects of misoperating various devices, vehicles, etc.

Some of the first studies performed were concerned with perception of vibration magnitude (acceleration), frequency dependence, exposure time (92-94), and the effects of seat belts and other body restraint systems (96). These led in part to the conclusion, that man's perceived indurance was both frequency-dependent and magnitude dependent, with the lowest tolerance being in the 4-8 Hz human whole-body resonance range (73,93-95) for motion (Z) sinusoidal vibration with seated subjects. Chaney, Parks and Snyder determined for short term exposures of a few minutes that acceleration levels at 0.5-1.5 g in the 4-8 Hz range could greatly alarm their subjects. When they reduced the acceleration to 0.02-0.05g, the alarming was barely perceptible in the same frequency range (96-98). Brumaghim (100) attempted to duplicate the aforementioned previous work (97) by using 17 Hz background vibration (0.38-0.68g) into the 4-8 Hz resonance range. The results indicated a masking effect due to the 17 Hz component, with less 4-8 Hz vibration required to obtain an annoyance response with increases of the 17 Hz component. In a later NIOSH study by Cohen, Wasserman, and Hornung (101) used six subjects to perform a four-limb coordination task atop a large vertical shaker table attempting to simulated a living heavy equipment driving environment. The test conditions were no vibration, 2.5 Hz sinusoidal vibration (0.07 g_{rms}), 5.0 Hz sinusoidal vibration (0.07 g_{rms}) and a limited spectrum containing mixture of 2.5 Hz and 5.0 Hz (totaling 0.07 g_{rms}). The purpose was to determine for 2.5 hour exposures the ability of these subjects to perform coordination tasks (response speed) under a limited spectral condition (e.g., mixture), versus the 5 Hz resonance condition, versus a 2.5 Hz nonresonance condition and then without vibration. The results indicated the worse performance response was under the mixture condition, followed by the 5 Hz, then 2.5 Hz, and finally no vibration conditions. These results seem to indicate that although much of the basic laboratory studies use discrete sinusoidal conditions, that when even a limited mixture of vibration is used (which begins to approximate the workplace), the performance results can be quite different than when discrete sinusoids are used above. Hornick (102,103) used discrete sinusoidal vibration at acceleration levels of 0.15, 0.25, 0.35 g in a frequency range of 0.9-6.5 Hz sequentially in the vertical (Z), horizontal (Y), and lateral (X) directions and found response time was unaffected by this vibration (93).

Shoenberger, however, has questioned Hornick's results, because Hornick used both a two dimensional tracking task and choice response time (104) whereas Shoenberger found increased response times at 5 and 8 Hz (Z direction) and at 1, 3, 5 and 8 Hz in the X direction, and 1 and 3 Hz in the Y direction.

Various tracking tasks have been studied with a progressive tracking task deterioration with increased vibration acceleration, usually in the human whole-body resonance range of 4-8 Hz (Z direction); for X and Y directions the greatest deterioration seems to occur at the lower frequencies 1, 3, 5 Hz (73,93). Huddleston (105,106) found certain mental tasks involving mental addition and recent memory were slower at 4.8, 6.7, 9.5, and 16 Hz range (all at 0.5 g) than in nonvibration control periods. Grether reports that tasks involving pattern recognition and visual monitoring seems not to be affected by vibration (107).

We have thus far addressed only linear vibration and not angular or rotational vibration (e.g., pitch, yaw and roll). In studies by Sjoflot and Suggs (108,109) where transverse angular vibration was compared to linear vertical vibration while attempting to simulate forms of farm tractor driving, these investigators found more degraded tracking task performance with angular vibration than just vertical vibration alone. The worst performance was a combination of angular and vertical vibration at 1, 1.7, 2.5, 24 Hz (all at 0.25 and 0.50 g).

What are the direct mechanical effects on performance of human subjects exposed to whole-body vibration? The key seems to be that performance decrement occurs mostly in the whole-body resonance frequencies ranges. Where resonance represents that frequency (or frequencies) where the body is optimally mechanically coupled (or tuned) to the vibration source. At resonance there is maximum mechanical vibration energy transfer between the vibration source and the body with an actual amplification of the incoming vibration by the body (see Chapter 3 for details). It would seem reasonable

TABLE 2.6
Resonance Frequencies of Various Body Parts of Man.
(from Reference 114)

Posture	Body part	Vibration direction	Range of resonance frequency
Lying	Foot	x	16 - 31 Hz
	Knee	x	4 - 8 Hz
	Abdomen	x	4 - 8 Hz
	Chest	x	6 - 12 Hz
	Bone of head	x	50 - 70 Hz
	Foot	y	0.8 - 3 Hz
	Abdomen	y	0.8 - 4 Hz
	Head	y	0.6 - 4 Hz
	Foot	z	1 - 3 Hz
	Abdomen	z	1.5 - 6 Hz
	Head	z	1 - 4 Hz
Standing	Knee	x	1 - 3 Hz
	Shoulder	x	1 - 2 Hz
	Head	x	1 - 2 Hz
	Whole body	z	4 - 7 Hz
Sitting	Trunk	z	3 - 6 Hz
	Chest	z	4 - 6 Hz
	Spine	z	3 - 5 Hz
	Shoulder	z	2 - 6 Hz
	Stomach	z	4 - 7 Hz
	Eyeballs	z	20 - 25 Hz

that performance decrements would be exacerbated in the resonance regions. Indeed, this does seem to be the case, for example, it is very difficult to maintain skilled performance in the 4-8 Hz (nominally 5 Hz) vertical range (73,93,94,101,110-113) whether the vibration be continuous or random. It would appear that decrements in tracking are worse in the vertical direction than in the other two dimensions (104). The addition of an arm rest in one study helped reduce tracking error under vibration (114) and may represent a pallative measure.

The NIOSH work simulation study already cited (101) and others (93) of long-term (e.g., multiple hour) vibration performance studies seem to indicate that vibration produces progressive increases in tracking error, but recovery after rest periods. It would appear that fatigue, vibration, and rest periods and how they are divided through the workday all play a complex role in overall performance decrement. There certainly needs to be more research done in the area of workplace performance under vibration conditions especially under resonance to clarify these issues. Finally, since resonance is such an important concept, Dupius (115) has attempted to tentatively summarize various body-part resonances (See Table 2.6) from several whole-body studies and is presented here for the reader's information.

REFERENCES

1 W. Taylor (Ed.), The Vibration Syndrome, Academic Press, London, 1974.
2 W. Taylor and P.L. Pelmear (Eds.), Vibration White Finger in Industry, Academic Press, London, 1975.
3 D.E. Wasserman and W. Taylor (Eds.), Proceedings of the International Occupational Hand-Arm Vibration Conference, DHEW/NIOSH Public. No. 77-170, 1977.
4 A.J. Brammer and W. Taylor (Eds.), Vibration Effects on the Hand and Arm in Industry, Wiley and Sons Publishers, New York, 1982.
5 M. Raynaud, Local Asphyxia and Symmetrical Gangrene at the Extremities, M.D. Thesis, Paris (1862). (Translated into English in: Selected Monographs.) London, New Sydenham Society, 1888.
6 G. Loriga, Pneumatic Tools: Occupation and Health, Ball. Inspet. Lorboro, 2 (1911) 35-37.
7 W. Taylor, A Century of Devotion to Industrial Health: A Tribute to One of America's Great Physicians - Dr. Alice Hamilton (1869-1970), Applied Industrial Hygiene (In Press).
8 E. Sicherman, A Life in Letters - Alice Hamilton, Harvard University Press, Cambridge, Mass., 1984.
9 A. Hamilton, A Study of Spastic Anemia in the Hands of Stonecutters: An Effect of the Air Hammer on the Hands of Stonecutters, Industrial Accidents and Hygiene Series, Bulletin 236, No. 19, U.S.D.O.L., Bureau of Labor Statistics, 1918.
10 V. Behrens, D. Wasserman, W. Taylor and T. Wilcox, Vibration Syndrome in Chipping and Grinding Workers, J. Occup. Medicine (Special Supplement), 26 (1984) 765-788.

11 M.D. Grounds, Raynaud's Phenomenon in Users of Chain Saws, Med. J. of Australia, 1 (1964) 270-272.
12 F. Huzl, R. Stolarik and J. Marnerova, Damage Due to Vibrations When Felling Timber by Power Saws, Pracov. Lek. 23 (1971) 7-15.
13 T. Miura, K. Kimura and V. Tominaga, On Raynaud's Phenomenon of Occupational Origin Due to Vibrating Tools - Its Incidence in Japan, Institute for Science of Labor, Report #65, 1966.
14 I. Pyykko, The Prevalence and Symptoms of Traumatic Vasospastic Disease Among Lumberjacks in Finland: A Field Study. Work-Environment-Health, 11 (1974) 118-131.
15 E.E. Dart, Effects of High Speed Vibrating Tools on Operators Engaged in the Airplane Industry, Occup. Medicine, 1 (1946) 515-550.
16 J.H. Hunt, Raynaud's Phenomenon in Workmen Using Vibrating Instruments, Proc. R. Soc. Med. 30 (1936) 171-178.
17 M.A.F. Hardgrove and N.W. Barker, Pneumatic Hammer Disease: A Vasospastic Disturbance of the Hands in Stonecutters, Proc. Staff Meeting of Mayo Clinic, 8 (1933) 345-349.
18 M. Seyring, Maladies from Work with Compressed Air Drills, Bull. Hyg., 6 (1931) 25.
19 F.J. Peters, A Disease Resulting from the Use of Pneumatic Tools, Occup. Med., 2 (1946) 55-66.
20 J.N. Agate and H.A. Druett, A Study of Portable Vibrating Tools in Relation to the Clinical Effects Which They Produce, Brit. J. Ind. Med., 4 (1947) 141-163.
21 J.N. Agate, H.A. Druett and J.B.L. Tombleson, Raynaud's Phenomenon in Grinders of Small Metal Castings, Brit. J. Ind. Medicine, 3 (1946) 167-170.
22 J.N. Agate, An Outbreak of Cases of Raynaud's Phenomenon of Occupational Origin, Brit. J. Ind. Med., 2 (1945) 10-16.
23 D.E. Wasserman, W. Taylor, V. Behrens, S. Samueloff and D. Reynolds, VWF Disease in U.S. Workers Using Pneumatic Hand-Tools (Vol. 1 - Epidemiology), DHHS/NIOSH Public. No. 82-118, 1982.
24 W. Taylor, T. Wilcox and D. Wasserman, Health Hazard Evaluation Report - Neenah Foundry Co., DHHS/NIOSH Public. No. HHE-80-189-870, 1981.
25 M. Bovenzi, L. Petronio and F. DiMarino, Epidemiological Survey of Shipyard Workers Exposed to Hand-Arm Vibration, Int. Arch. Occup. Envir. Health, 46 (1980) 251-256.
26 D.E. Wasserman, D. Reynolds, V. Behrens, W. Taylor, S. Samueloff and R. Basel, VWF Disease in U.S. Workers Using Pneumatic Hand-Tools (Vol. 2 - Engineering), DHHS/NIOSH Public. No. 82-101, 1982.
27 D.D. Reynolds, R. Basel, D.E. Wasserman and W. Taylor, A Study of Hand Vibration on Chipping and Grinding Operations: Vibration Acceleration Levels Measured on Pneumatic Tools Used in Chipping and Grinding Operations. J. Sound and Vib., 95 (1984) 479-497.
28 L.J. Pecora, M. Udel and R.P. Christman, Survey of Current Status of Raynaud's Phenomenon of Occupational Origin, J. Am. Ind. Hyg. Assoc., 21 (1960) 80-83.
29 N. Williams and E.B. Byrne, An Investigation to Determine the Qualitative and Quantitative Extent of Health Records and Injury Claims: Records of Workers Exposed to Hand Tool Vibration in U.S. Industry, Final Report, NIOSH Contract HSM-99-73-56, 1974.
30 W. Taylor, D. Wasserman, V. Behrens, D. Reynolds and S. Samueloff, Effect of the Air Hammer on the Hands of Stonecutters. The Limestone Quarries of Bedford, Indiana, Revisited, Brit. J. of Ind. Medicine, 41 (1984) 289-295.
31 I. Pyykko and J. Hyvarinen, The Physiological Basis of the Traumatic Vasospastic Disease: A Sympathetic Vasoconstrictor Reflex Triggered by High Frequency Vibration? Work-Environ-Health, 10 (1973) 36-47.

32 W.F. Ashe, W.T. Cook and J.W. Old, Raynaud's Phenomenon of Occupational Origin, Arch. Environ. Health, 5 (1962) 333-343.
33 T. Azuma, T. Ohhashi, and M. Sakaguchi, Vibration-Induced Hyperresponsiveness of Arterial Smooth Muscle to Noradrenaline With Special Reference to Raynaud's Phenomenon in Vibration Disease. Cardiovasc. Res., 12 (1978) 758-764.
34 P. Pellegrini, G. Martines and C. Longhini, A Study of the Anatomical Substrate of Occupational Microangiopathy Caused by Vibrating Instruments, Med. Lavoro, 59 (1968) 180-208.
35 R.M. Nerem, Vibration-Induced Arterial Shear Stress: The Relationship to Raynaud's Phenomenon of Occupational Origin, Arch Environ. Health., 26 (1973) 105-110.
36 W.F. Ashe and N. Williams, Occupational Raynaud's, Arch. Environ. Health, 9 (1964) 425-433.
37 R. Blunt, A. George and R. Hurloro, Hyperviscosity and Thrombotic Changes in Idiopathic and Secondary Raynaud's Syndrome, Br. J. Haematol. 45 (1980) 651-658.
38 K.B. Goyle and J.A. Dormandy, Abnormal blood viscosity in Raynaud's Phenomenon, Lancet, 3 (1976) 1317-1318.
39 D. Wasserman, W. Carlson, S. Samueloff, W. Asburry and T. Doyle, A Versatile Simultaneous Multifinger Photocell Plethysmorgraphy System for Use in Clinical and Occupational Medicine, J. Med. Inst. 13 (1979) 232-234.
40 S. Samueloff, R. Miday, D. Wasserman and V. Behrens, Peripheral Vascular Insufficiency Test Using Photocell Plethysmography, J. Occup. Med., 23 (1981) 643-646.
41. S.L. Nielsen, Raynaud's Phenomena and Finger Systolic Pressure During Cooling, Scand. J. Clin. Lab. Invest., 38 (1978) 765-770.
42 S. Renfrew, Finger Tip Sensation: A Routine Neurologist Test, Lancet, 1 (1969) 396-397.
43 W.S. Carlson, S. Samueloff, W. Taylor and D. Wasserman, Instrumentation for Measurement of Sensory Loss in the Fingertips, J. Occup. Med., 21 (1979) 260-264.
44 H. Streeter, Tactility Measurement Instrumentation, J. Amer. Ind. Hygiene Assoc., 31 (1970) 87-91.
45 A.M. Seppalanien, Neurophysiological Detection of Vibration Syndrome in the Shipbuilding Industry, In Vibration and Work (O. Kerhonen, editor), Inst. of Occupational Health, 1976, pp, 63-71.
46 E.N. Corlett, N.K. Akinmayowa and K. Sivayoganathan, A New Aesthesiometer for Investigating Vibration White Finger (VWF), Ergonomics, 24 (1981) 49-54.
47 W. Taylor and D. Wasserman, Occupational Vibration, In Occupational Medicine (C. Zenz, Editor), Yearbook Medical Publishers, Chicago, (In Press)
48 G. Talpos, M. Horrocks and J.M. White, Plasmaphoris in Raynaud's Disease, Lancet, 1 (1978) 416-417.
49 L.L. Peterson, R.J. Rodeheffer and R.S. Surrit, Raynaud's Syndrome: Can We Finally Control the Vasospasm? Data Centrum, 1 (1984) 11-16.
50 J.M. Porter and J.M. Edwards, Raynaud's Syndrome - Underlying Mechanisms, Drug Therapy, 10 (1985) 72-90.
51 D.S. Olton and A.R. Noonberg, Biofeedback - Clinical Applications in Behavioral Medicine, Prentice-Hall Publishers, Englewood Cliffs, N.J., 1980.
52 R.R. Freedman, P. Ianni and P. Weing, Behavioral Treatment of Raynaud's Phenomenon in Scleroderma, J. Behavioral Medicine, 7 (1984) 343-353.
53 G.W. Thorn, R.D. Adams, E. Braunwald, K.J. Isselbacker and R.G. Petersdorf, Harrison's Principles of Internal Medicine, McGraw-Hill Publishers, New York, N.Y., 1977.

54 M. Mumemthaler, Neurology, Yearbook Medical Publishers, Chicago, Ill., 1977.
55 E. Tichauer and H. Gage, Ergonomic Principles Basic to Hand Tool Design, J. Amer. Ind. Hygiene Assoc., 38 (1977) 622-623.
56 S. Rothfleish and D. Sherman, Carpal Tunnel Syndrome: Biomechanical Aspects of Occupational Occurrence and Implications Regarding Surgical Management, Orthop. Review, 7 (1978) 107-109.
57 T. Armstrong and D. Chaffin, Carpal Tunnel Syndrome and Selected Personal Attributes, J. Occup. Med., 21 (1979) 481-486.
58 L.S. Cannon, E.J. Bernacki and S.D. Walter, Personal and Occupational Factors Associated with Carpal Tunnel Syndrome, 23 (1981) 255-258.
59 A.M. Seppalainen, Private Communication with the Author, 1983.
60 D.E. Wasserman and D. Badger, The NIOSH Plan for Developing Industrial Vibration Exposure Criteria, J. Safety Research, 4 (1973) 146-154.
61 H.H. Cohen, D.E. Wasserman and R. Hornung, Human Performance and Transmissibility Under Sinusoidal and Mixed Vertical Vibration, Ergonomics, 29 (1977) 207-216.
62 G.J. Gruber, Relationship Between Whole-Body Vibration and Morbidity Patterns Among Interstate Truck Drivers, U.S. DHEW/NIOSH Public. No. 77-167, 1976.
63 G.J. Gruber and H.H. Ziperman, Relationship Between Whole-Body Vibration and Morbidity Patterns Among Motor Coach Operators, U.S. DHEW/NIOSH Public. No. 75-104, 1974.
64 T.H. MIlby and R.C Spear, Relationship Between Whole-Body Vibration and Morbidity Patterns Among Heavy Equipment Operators, U.S. DHEW/NIOSH Public. No. 74-131, 1974.
65 R.C. Spear, C.A. Keller, V. Behrens, M. Hudes and D. Tarter, Morbidity Patterns Among Heavy Equipment Operators Exposed to Whole-Body Vibration: Follow-up to a 1974 Study, U.S. DHEW/NIOSH Public. No. 77-120, 1976.
66 R. Rosegger and S. Rosegger, Health Effects of Tractor Driving, J. Agric. Engr. Res., 5 (1960) 241-275.
67 H. Seris and R. Auffret, Measurement of Low Frequency Vibration in Big Helicopters and Their Transmission to the Pilot, NASA Technical Translation Report NASA-TTF-471, 1967.
68 V. Behrens, Analysis of Results from U.S. Army Survey of Back Discomfort in Rotary Wing Aviators (Interagency Agreement Between NIOSH and U.S. Army), DHHS/NIOSH/DBBS/PAEB Internal Memorandum Report, Cincinnati, Ohio, April 11, 1983.
69 D.F. Shanahan and T.E. Reading, Helicopter Pilot Back Pain: A Preliminary Study, Aviation Space and Envir. Medicine, 55 (1984), 117-121.
70 J.G. Fitzgerald and J. Crotty, The Incidence of Backache Among Aircrew and Ground Crew in the Royal Air Force, Report FPRC/1313, Flying Personnel Research Committee, 1972.
71 J.L. Kelsey, An Epidemiological Study of the Relationship Between Occupations and Acute Herniated Lumbar Intervertebral Discs, Int. J. Epid., 4 (1975) 197-205.
72 J.L. Kelsy and R.J. Hardy, Driving of Motor Vehicles as a Risk Factor for Acute Herniated Intervertebral Discs, Amer. J. Epid., 102 (1975) 63-73.
73 J.C. Guignard and P.F. King, Aeromedical Aspects of Vibration and Noise, AGARDograph No. 17, NATO, Technical Editing and Reproduction Ltd, Charlotte St. London, 1972.
74 W.B. Hood, R.M. Murray, C.W. Unschel, J.A. Bowers and J.G. Clark, Cardiopulmonary Effects of Whole-Body Vibration in Men, J. Appl. Physiol., 21 (1966) 1725-1731.
75 J.H. Dines, J.H. Sutphen, L.B. Roberts and W.F.Ashe, Intravascular Pressure Measurements During Vibration, Arch. Envir. Health, 11 (1965) 323-362.

76 V. Guillemin and P. Wechsberg, Physiological Effects of Long Term Repetitive Exposure to Mechanical Vibration, J. Aviation Med., 24 (1953) 208-221.
77 R. Coermann, Investigations into the Effects of Vibration on the Human Body, Luftfahrt Medizin., 4 (1940) 73-117.
78 L.R. Duffner, L.M. Hamilton and M.A. Schmitz, Effects of Whole-Body Vertical Vibration on Respiration in Human Subjects, J. Appl. Physiol., 17 (1962) 913-916.
79 W.F. Ashe, Physiological and Pathological Effects of Mechanical Vibration on Animals and Man, Ohio State University Research Foundation, Report 862-4, Columbus, Ohio 1961.
80 J.C. Guignard, Vibration, A Textbook of Aviation Physiology, Peragomen Press, Oxford, 1965.
81 J.C. Guignard, Chapter 15 - Vibration, Patty's Industrial Hygiene and Toxicology, 2nd. Edition, John Wiley and Sons Publishers, New York, N.Y., 1985.
82 R. Jakubowski, General Characteristics of Vibration at Various Workplaces in Agriculture and Forestry, Med. Wicj., 4 (1969) 47-50.
83 D.V. Sturges, D. Badger, R. Slarve and D.E. Wasserman, Laboratory Studies on Chronic Effects of Vibration Exposure, Proc. AGARD Conf. on Vibration and Combined Stresses in Advanced Systems, Paper BIO-1, AGARD-CPP-145, Oslo, Norway, April, 1974.
84 D. Badger, D. Sturges, R. Slarve and D.E. Wasserman, Serum and Urine Changes in Macaca Mulatta Following Prolonged Exposures to 12 Hz, 1.5 g Vibration, Paper B11-1 (Ibid).
85 D.E. Wasserman and D.W. Badger, Vibration and the Worker's Health and Safety, DHEW/NIOSH Technical Public. No. 77, 1973.
86 A.J. Benson, Possible Mechanisms in Motion and Space Sickness, Proc. of the European Symposium on Life Sciences Research in Space, 101-108, Cologne, Germany, 1977.
87 S.R. Olsen, Human Effectiveness in the Ship Motion Environment, Center for Naval Analysis, (CNA) 77-0665, Arlington, VA., 1977.
88 H.I. Chinn and P.K. Smith, Motion Sickness, Pharmacol. Rev., 7 (1955) 33-82.
89 G.H. Crampton, Studies of Motion Sickness. Physiological Changes Accompanying Sickness in Man, J. Appl. Physiol., 7 (1955) 501-507.
90 E.V. Dahl, J.J. Franks, J.R. Prigmore and R.L. Cramer, Adrenal Cortical Response in Motion Sickness, Arch. Environ. Health, 7 (1963) 86-91.
91 K.E. Money, Motion Sickness, Physiol. Rev., 50 (1970), 1-39.
92 D.E. Wasserman, T.E. Doyle and W.C. Asburry, Whole-Body Vibration Exposure of Workers During Heavy Equipment Operation, DHEW/NIOSH Public. No. 78-53, 1978.
93 R.W. Shoenberger, Human Response to Whole-Body Vibration, Perceptual and Motor Skills, Monograph Supplement 1-V34, Missoula, Montana (Same as AMRL Report TR-71-68), 1972.
94 E.B. Magid, R.R. Coermann and G.H. Ziegenruecker, Human Tolerance to Whole-Body Sinusoidal Vibration, Aerospace Medicine, 31 (1960) 915-924.
95 M.J. Mandel and R.D. Lowry, One Minute Tolerance in Man to Vertical Sinusoidal Vibration in the Sitting Position, AMRL 62-121, 1962.
96 W.E. Temple, N.P. Clarke, J.W. Brinkley and M.J. Mandel, Man's Short-Time Tolerance to Sinusoidal Vibration, Aerospace Medicine, 35 (1964) 923-930.
97 R.E. Chaney and D.L. Parks, Visual-Motor Performance During Whole-Body Vibration, Boeing Co. Tech. Report D3-3512-5, Wichita, Kansas, 1964.
98 D.L. Parks, Defining Human Reaction to Whole-Body Vibration, Human Factors, 4 (1962) 305-314.
99 D.L. Parks and F.W. Snyder, Human Reaction to Low Frequency Vibration, Boeing Co. Tech. Report D3-3512-1, Wichita, Kansas, 1961.
100 S.H. Brumagheim, Subjective Reaction to Dual Frequency Vibration, Boeing Co. Tech. Report D3-7562, Wichita, Kansas, 1967.

101 H.H. Cohen, D.E. Wasserman and R.W. Hornung, Human Peformance and Transmissibility Under Sinusoidal and Mixed Vertical Vibration, Ergonomics, 20 (1977) 207-216.
102 R.J. Hornick, Problems in Vibration Research, Human Factors, 8 (1966) 481-492.
103 R.J. Hornick, Effects of Whole-Body Vibration in Three Directions Upon Human Performance, J. of Engr. Psychology, 1 (1962) 93-101.
104 R.W. Shoenberger, Human Performances as a Function of Direction and Frequency of Whole-Body Vibration, AMRL TR70-7, 1970.
105 H.F. Huddleston, Human Performance and Behavior in Vertical Sinusoidal Vibration, Institute of Aviation Medicine Report No. 303, Farnborough, U.K., 1964.
106 H.F. Huddleston, Effects of 4.8 and 6.7 Hz vertical Vibration on Handwriting and a Complex Mental Task with and Without Abdominal Restraint, Institute of Aviation Medicine Report No. 60, Farnborough, U.K., 1965.
107 W.F. Grether, Vibration and Human Peformance, Human Factors, 13 (1971) 203-205.
108 L. Sjoflot and C.W. Suggs, Low Frequency Angular Vibrations in the Roll Mode on Farm Tractors, J. of Agricult. Engr. Research, 17 (1972) 22-32.
109 L. Sjoflot and C.W. Suggs, Human Reactions to Whole-Body Transverse Angular Vibrations Compared to Linear Vertical Vibrations, Ergonomics, 16 (1973) 455-468.
110 M.A. Larne, The Effects of Vibration on Accuracy of a Positioning Tool, J. of Envir. Sciences, 8 (1965) 33-35.
111 D.E. Goldman and H.E. VonGierke, The Effects of Shock and Vibration on Man, U.S. Naval Medical Research Institute Report No. 60-3, Bethesda, Maryland, 1960.
112 J.C. Guignard and A. Irving, Effects of Low Frequency Vibration on Man, Engineering, 190 (1960) 364-367.
113 R.D. Dean, R.J. Farrell and J.D. Mitt, Effect of Vibration on the Operation of Decimal Input Devices, Human Factors, 11 (1969) 257-272.
114 G. Torle, Tracking Peformance Under Random Acceleration: Effects of Control Dynamics, Ergonomics, 8 (1965) 481-486.
115 H. Dupuis, Data on Acceleration Versus Frequency with Respect to Preservation of Health, ISO/TC108/SC4/WG2, Working Document 2-N-128, April, 1984 (unpublished).

SUPPLEMENTARY BIBLIOGRAPHY

Hand-Arm Vibration

K. Ikehata, S. Kawauchi, F. Kohno, M. Nishiyama and N. Ide, Increased Platelet Function and Von Wittenbrand Factor in Vibration Syndrome, Tokushima J. Exp. Med. 27 (1980) 23-28.

D. Knapikova, Serum Proteins in Vibration Disease, Medycyna Pracz, 21 (1970) 307-309.

T. Petelnc, A. Kujawska, A. Misiewica, J. Strudewski, J. Kusniarz, Clinical and Occupational Problems of Vibration Disease, Med. Pracz, 4 (1966) 296-298.

D. Leys, Diffuse Scleroderma and Raynaud's Phenomenon, Lancet, 2 (1939) 692.

J.F. Brailsford, Pathological Changes in Bones and Joines Induced by Injury, Brit. Med. J., 2 (1936) 657-663.

N. Williams, Biological Effects of Segmental Vibration, J. Occup. Med., 17 (1975) 37-39.

T. Kakosy and L. Szepesi, Effects of vibration Exposure on the Localization of Raynaud's Phenomenon in Chain Saw Operators, Work-Envir.-Health, 10 (1973) 134, 139.

M. Bovenzi, A. Fiorito, C. Giansante, J. Calabrese and C. Negro, Platelet Function and CLotting Parameters of Vibration-Exposed Foundary Workers, Scand. J. Work-Envir.-Health, 9 (1983) 347-352.

K. Suzuki, M. Ijichi, T. Matsuki, A. Seki, H. Tanaka, M. Ogata and K. Noro, The Hand and Environment - Physiological Changes of Skin Temperature, Vibration Sensibility, Pinch Strength of the Hand, and Oxygen Uptake in Two Climatic Conditions, J. UOEH, 3 (1981) 109-115, (Japan).

J.E. Thomas, E.H. Lambert, and K.A. Cseuz, Electrodiagnostic Aspects of the Carpal Tunnel Syndrome, Arch. Neurol., 16 (1967) 635-641.

D.S. Chatterjee, D.D. Barwick and A. Petrie, Exploratory Electromyography in the Study of Vibration-Induced White Finger in Rock Drillers, Brit. J. Ind. Med., 39 (1982) 89-97.

A.M. Seppalainen, Nerve Conduction in the Vibration Syndrome, Work-Envir.-Health, 7 (1970) 82-84.

R.P. Jepsen, Raynaud's Phenomenon - A Review of the Clinical Problem, Ann. R. Coll. Surg., 9 (1951) 35-51.

B. Hellstrom and K. Myhre, A Comparison of Some Methods of Diagnosing Raynaud's Phenomenon of Occupational Origin, Br. J. Ind. Med., 28 (1971) 272-279.

K. Sedlacek, Biofeedback for Raynaud's Disease, Psychosomatosis, 8 (1979) 538-547.

E.V. Allen and G.E. Brown, Raynaud's Disease: A Critical Review of Minimal Requisites for Diagnosis, Amer. J. Med. Sciences, 183 (1932) 187-200.

E.E. Velayos, Clinical Correlation Analysis of 137 Patients with Raynaud's Phenomenon, Amer. J. Med. Sciences, 262 (1971) 347-349.

E. Lukas and V. Kuzel, Clinical and Electromyographic Diagnosis of Damage to the Peripheral Nervous System Due to Local Vibrations, Internatl. Archiv. fur Arbeitsmedizin, 28 (1971) 239-249.

N. Olsen and S.L. Nielsen, Diagnosis of Raynaud's Phenomenon in Quarrymen's Traumatic Vasospastic Disease, Scand. J. Work Envir. and Health, 5 (1979) 249-256.

J.T. Sappington, E.M. Fiorito and K.A. Brehony, Biofeedback as Therapy in Raynaud's Disease, Biofeedback and Self Regulation, 4 (1979) 155-169.

H.A. Kontas and A.J. Wasserman, Effect of Reserpine in Raynaud's Phenomonen, Circulation, 29 (1969) 259-266.

J.D. Coffman and W.T. Davies, Vasospastic Diseases: A Review, Progress in Cardiovascular Diseases, 18 (1975) 123-146.

B.E. Petrenko and M.L. Vydin, Differential Diagnosis of Peripheral Vasomotor Disorders Produced by Vibration and Other Causes, Noise and Vibration Bulletin, 19-22, February, 1983.

J.E. Hansson, L. Eklund, S. Kihlberg, A. Kjellberg, I. Sternerup, A. Utter, K. Weman and C.E. Ostergren, Exposure to Vibration in Car Repair Work - Comparison of Tools and Methods, Arbete Och Halsa, 3 (1985) 1-48

R. Dandanell and K. Engstrom, Vibrations from Percussive Tools such as Riveting Tools in the Frequency Range 6 Hz to 10 mHz and Raynaud's Phenomenon, Proc. of the Fourth International Symposium on hand-Arm Vibration, Helsinki, Finland, May, 1985.

D.E. Wasserman, Raynaud's Phenomenon as it Relates to Hand-Tool Vibration in the Workplace, J. Amer. Ind. Hygiene Assoc., 46 (1985) 10-18.

Whole-Body Vibration

H.R. Jex and J.W. Zellner, Significant Factors in Truck Ride Quality, U.S. Fed. Hwy. Admin. Report No. FHWA/RD-81/139, 1981.

R.W. Shoenberger and C.S. Harris, Psychophysical Assessment of Whole-Body Vibration, Human Factors, 13 (1971) 41-50.

R.W. Shoenberger, Comparison of the Subjective Intensity of Sinusoidal, Multifrequency and Random Whole-Body Vibration, Aviation, Space and Environmental Medicine, 47 (1976) 856-862.

R.W. Shoenberger, An Investigation of Human Information Processing During Whole-Body Vibration, Aerospace Medicine, 45 (1974) 143-153.

H.C. Sommer and C.S. Harris, Combined Effects of Noise and Vibration on mental Peformance as a Function of Time of Day, Aerospace Medicine, 43 (1972) 479-482.

T. Miwa, Thresholds of Perception of Vibration in Recumbent Man, J. Acoustical Society of America, 75 (1984) 849-854.

J. Mathews, Ride Comfort for Tractor Operators: Review of Existing Information, J. Agric. Engr. Res., 9 (1964) 3-31.

'D. Dieckmann, A Study of the Influence of Vibration on Man, Ergonomics, 1 (1958) 347-355.

F. Pradko and R.A. Lee, Vibration Comfort Criteria, Society of Automotive Engrs. Paper 660139, New York, 1966.

W.F. Grether, Effects on Human Peformance of Combined Environmental Stresses, AMRL TR-70-68, 1970.

D.H. Drazin, Factors Affecting Vision During Vibration, Research 15 (1962) 275-280.

R.A. Dudek and D.E. Clemens, Effect of Vibration on Certain Psychomotor Responses, J. Engr. Psychology, 4 (1965) 127-143.

Chapter 3

Mechanical Vibration Fundamentals

3.1 INTRODUCTION

In this chapter we present some basic mechanical vibration concepts necessary for an understanding of physical measurements and their relationship to the human body.

3.2 MECHANICAL VIBRATION (1-4)

3.2.1 Periodic motion

Mechanical vibration is a vector quantity and thus is defined by a magnitude and a direction. The motion may be periodic, or random. Periodic vibration refers to an oscillating motion of a body, about a reference position. The motion then repeats itself exactly after a certain time period; this is referred to as the frequency of vibration. The simplest form of periodic vibration is simple harmonic motion which when plotted as a function of time, is represented by a sinusoidal curve, Fig. 3.1a. T is the time period of vibration, namely, the time elapsed between two successive, conditions of the motion.

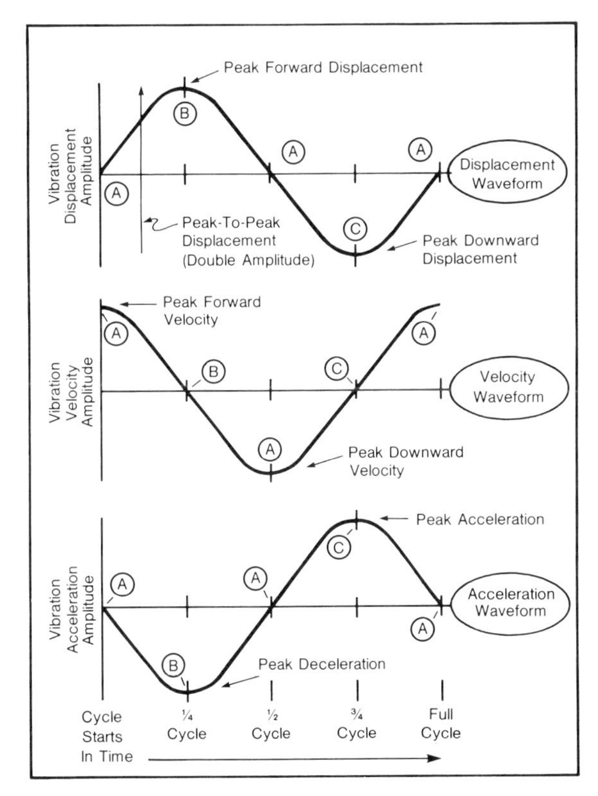

Fig. 3.1a Displacement, velocity, and acceleration waveforms and their phase relationships.

The relationship between time "t" (in seconds) and frequency "f" (in Hertz) is simply:

$$f = \frac{1}{T} \qquad (1)$$

If the vibration has the form of a pure translational motion along a given axis (x), the instantaneous displacement of the moving body from the reference position can be mathematically described by the equation:

$$X = X_{peak}\sin(2\pi\frac{t}{T}) = X_{peak}\sin(2\pi ft) = X_{peak}\sin(wt) \qquad (2)$$

where: $w = 2\pi f$ = angular frequency

X_{peak} = Maximum displacement from the reference position

t = time

The velocity (v) is the speed of a moving body which is mathmatically defined as the time rate of change of the displacement, or its first derivative, and is expressed in ft/sec. or m/sec. and is given by:

$$V = \frac{dx}{dt} = wX_{peak}(\cos wt) = V_{peak}\cos(wt) \qquad (3)$$

The acceleration "a" of a moving body is its changing speed, or the time rate of change of the velocity which is the second derivative of displacement. Acceleration is expressed in gravitational "g" units or meters/sec^2, (where 1g = 9.81 m/sec^2) and is given by:

$$a = \frac{d^2X}{dt^2} = \frac{dv}{dt} = -w^2X_{peak}\sin(wt) = -A_{peak}\sin(wt) = -A_{peak}(\text{displacement}) \qquad (4)$$

If we substitute $w = 2\pi f$ into equation (4), we obtain:

$$a = -4\pi^2f^2\sin(2\pi ft) \qquad (5)$$

Equation (5) can be used directly in terms of gravitational "g" units if we first divide it by g = 32.17 ft/sec^2 which is the acceleration due to gravity and then convert it into inches/sec^2 or (32.17) (12)=386.1. Then equation (5) becomes

$$a = -\frac{4\pi^2f^2X_{peak}\sin(wt)}{386.1} \qquad (6)$$

but peak refers to peak-to-peak or double amplitude (D.A.) thus we need to divide equation (6) by 2 since the motion of the body is either positive or negative, but never both occurring at the same time. We can also drop the sin (wt) and minus terms since we only seek peak acceleration. Thus, equation (6) becomes acceleration in "g" units.

$$A_{peak} = \frac{4\pi^2 f^2}{3.861} \frac{X}{2} = 0.51f^2x \qquad (7)$$

As an example, let us assume an object is moving at a vibration frequency of 12Hz, with a peak to peak or double amplitude displacement of 2 inches, and we seek its peak acceleration, then:

$$A_{peak} = .051(12)^2(2) = 14.68g$$

By elementary algebra, equation (7) can be easily solved for frequency:

$$f = \left(\frac{A}{.051X}\right)^{½} \qquad (8)$$

As an example, given A = 50g, X = 500 microinches peak-to-peak displacement = 500×10^{-6} inches. Solve for frequency f, we get:

$$f = \left[\frac{50}{(.051)(500\text{x}10^{-6})}\right]^{½} = \left(\frac{1}{.51\text{x}10^{-6}}\right)^{½} = \frac{10^3}{.714} = 1400 \text{ Hz}$$

Displacement is given by:

$$X = \frac{A}{.051f^2} \qquad (9)$$

when acceleration and vibration frequency are both known.

A simplified form of the velocity equation is:

$$V = wX_{peak} = 2\pi fX_{peak} \qquad (10)$$

but once again Xpeak refers to peak-to-peak (or Double) amplitude and we need to divide the equation by 2 or:

$$\text{peak velocity} = V_{peak} = \pi fx = 3.14fx \qquad (11)$$

with f(Hz) and X(inches), velocity is expressed in inches/sec.

$$f = \frac{V_{peak}}{3.14X} \qquad (12a)$$

Equation (12) is used when both velocity and displacement are known and frequency is unknown. Similarly, when displacement is unknown, with known velocity and frequency we get:

$$X = \frac{V_{peak}}{3.14f} \tag{12b}$$

As an example, given f = 35Hz, v=11 in/sec. (linear velocity) then X=0.10 inch linear peak-to-peak displacement.

Two additional equations are given next when displacement is given in Double Amplitude (D.A.) centimeters, and we seek acceleration in g's, or velocity in cm/sec.

$$\text{Acceleration (g's)} = (.02)\ (\text{D.A. cm})\ f^2(\text{Hz}) \tag{13}$$

$$\text{Velocity (cm/sec.)} = (\pi)\ (\text{D.A. cm})\ f(\text{Hz}) \tag{14}$$

From equations (3,4,5) it can be seen that the velocity leads the displacement by a phase angle of 90° and the acceleration leads the velocity by a phase angle of 90°. With Xpeak, Vpeak, and Apeak considered characterizing values for the respective magnitudes. Figure 3.1b graphically illustrates the relationships of phase and amplitude for displacement, velocity, and acceleration.

For completeness we mention a quantity only occasionally used in occupational vibration work, jerk, which is the time rate of change of acceleration. Jerk is sometimes used in subjective response studies and is discussed in Chapter 6.

It is important for the reader to take special note of equation (4) where acceleration is proportional to w^2 and hence frequency f^2. What this means in practical terms in the workplace is that merely viewing periodic motion of a moving object tells one little about its acceleration unless one knows the vibration frequency. To illustrate this point, at a frequency at 2Hz, 5 inches Double-Amplitude (DA) of displacement is required to produce 1g acceleration; at 20Hz, only 0.5 inches (DA) displacement will produce 1g acceleration; at 200 Hz, only 0.05 inches (DA) displacement will produce 1g acceleration. Thus as one observes a vibrating hand tool with the eye, for example, and there "appears to be" little displacement, there indeed can be many g's of acceleration (because of high frequency vibration components) impinging on workers hands. Similarly, observing a heavy equipment earth mover there will "appear to be" considerable displacement. Yet indeed there may be very little acceleration impinging on the driver, because of the

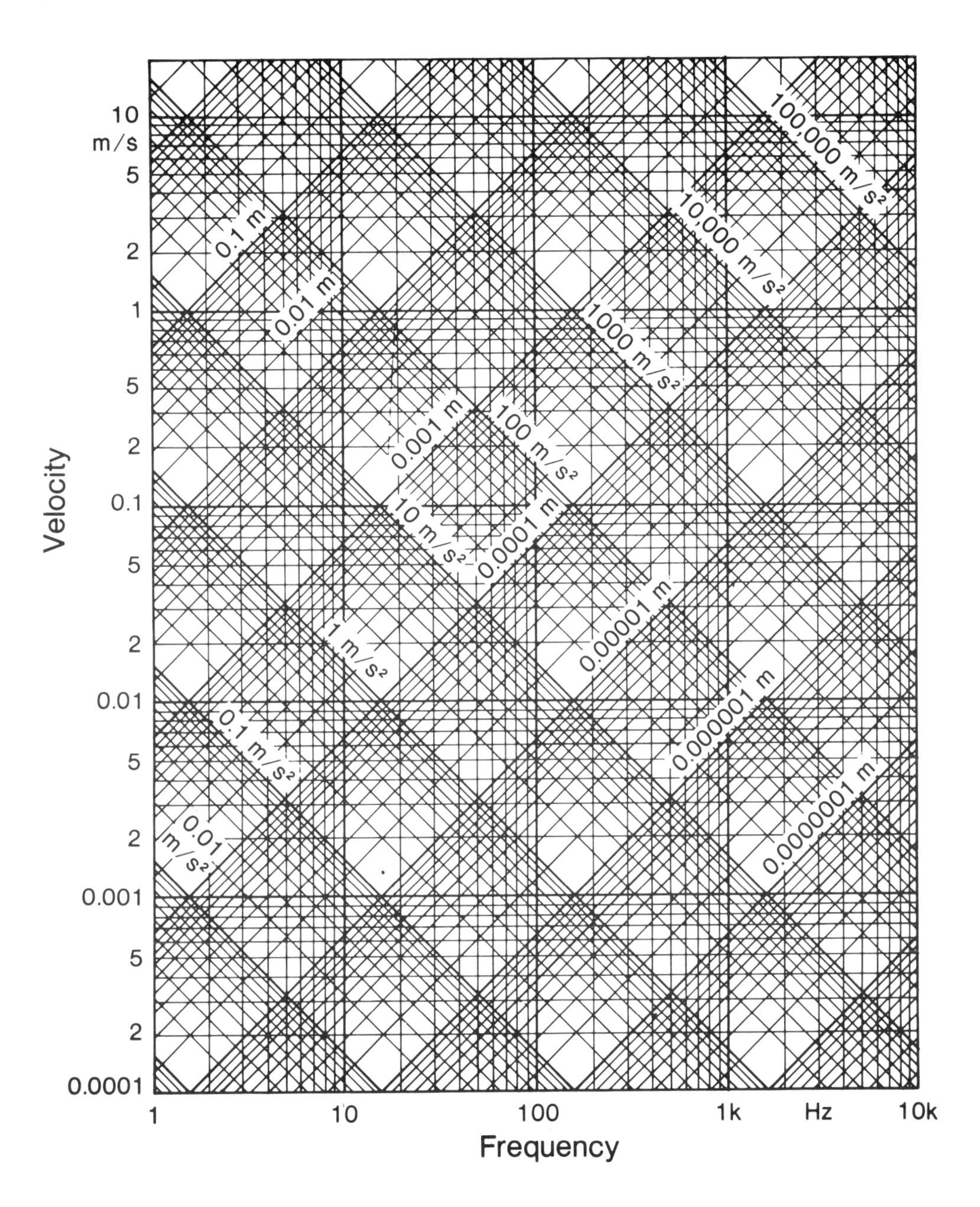

Fig. 3.1b Frequency, acceleration, velocity, displacement nomograph (RMS values). (Courtesy B & K Co.)

predominent low frequency vibration components of the motion. The important point here is simply: merely observing vibration with the eye, without measuring it, and not knowing the frequency content of the vibration and its acceleration, tells one little about the observed vibration.

(If the reader wishes to avoid mathematics, Figure 3.1 can be used to graphically determine acceleration, velocity, and displacement as a function of frequency).

Peak values are quite useful as long as pure sinusoidal or harmonic vibration is considered. But more complex periodic vibration contains multiple vibration frequencies, thus other descriptive quantities are necessary.

The absolute average value (avg) is used when the time history of vibration needs to be taken into account:

$$A_{avg} = \frac{1}{T} \int_{o}^{T} a(t)dt \qquad (15)$$

Equation (15) is simply summing the acceleration (or velocity, or displacement) values over the total measurement time period and then dividing the resultant by T.

A commonly used and more useful quantity is the "root-mean-square" (rms) value which is given by:

$$a_{rms} = \left[\frac{1}{T} \int_{o}^{T} a^2(t)dt\right]^{1/2} \qquad (16)$$

Equation (16) is an extension of equation (15), in this case the acceleration measurements are squared (thus negative acceleration values become positive values), next all of these values are summed over the measurement period. The sum is then divided by the total measurement time T. Finally the square root of the resulting value yields the rms answer. The importance of the rms value as a descriptive quantity is its direct relationship to the energy content of the vibration being measured.

For pure harmonic (sinusoidal) motion these relationships are shown in Figure 3.2 and are given as:

$$a_{rms} = \frac{\pi}{2\sqrt{2}} \; a_{avg} = \frac{1}{2\sqrt{2}} \; A_{peak} \qquad (17)$$

Nonsinusoidal periodic vibration have different waveshapes and thus there is a need to quantitatively describe them too. These are accomplished using the next two equations called "form-factor" (F_f) and "crest-factor," (Fc) respectively,

$$a_{rms} = F_f a_{avg} = \frac{1}{F_c} A_{peak} \quad (18)$$

$$F_f = \frac{a_{rms}}{a_{avg}} \quad (19)$$

$$F_c = \frac{a_{peak}}{a_{rms}} \quad (20)$$

For sinusoidal motion F_f and F_c become:

$$F_f = \frac{\pi}{2\sqrt{2}} = 1.11 \quad (21)$$

$$F_c = \sqrt{2} = 1.414 \quad (22)$$

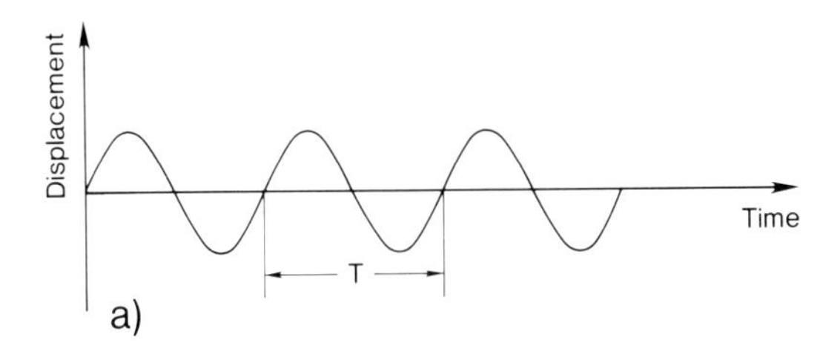

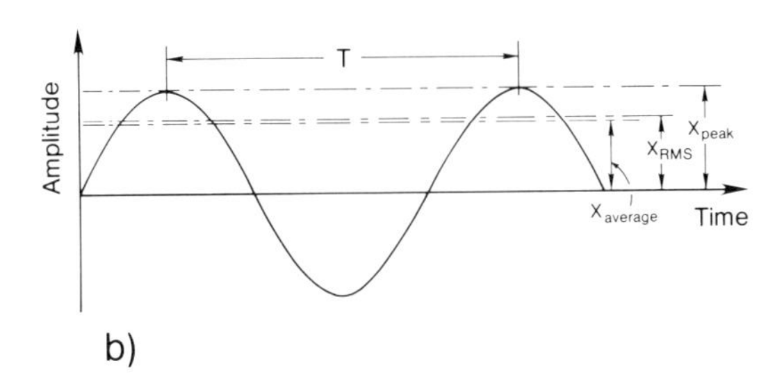

Fig. 3.2 (a) Diagram showing sinusoidal displacement vs. time, (b) relationships between peak, rms, and average values for sinusoidal vibration.

3.3 VIBRATION SPECTRUM ANALYSIS

In the previous section the need for a knowledge of the frequency content of the vibration being measured was indicated. Not only is it necessary to relate vibration frequency to acceleration, but just as important is that most mechanical vibration found in the workplace is not simply sinusoidal or pure harmonic motion. Much of it may be characterized as "complex periodic motion" containing many vibration frequencies all impinging on the worker at once. A method for describing this type of vibration is known as Fourier spectrum analysis. In the 1800's a French mathematician named Fourier showed that a complex periodic waveform was composed of a series of sinusoids of varying frequencies and amplitudes as given by:

$$F(t) = a_0 + a_1\sin wt + a_2\sin 2wt + a_3\sin 3wt + \cdot\cdot\cdot + a_n\sin(n)wt + b_1\cos wt + b_2\cos 2wt + b_3\cos 3wt + \cdot\cdot\cdot + b_n\cos(n)wt \quad (23)$$

The "a" and "b" values denote the respective amplitudes of each sinusoid at specific frequencies which compose the spectrum. The lone a_o term in the series depicts the DC or Zero Hz term which in most cases is not present and hence becomes zero. Figure 3.3 graphically displays the Fourier concept. Figure 3.3a depicts the measured resultant acceleration waveform (complex periodic). Figure 3.3b shows that there are actually two sinusoids (dashed lined sinusoid and low amplitude solid sinusoid) which by linear superposition algebraically sum to form the resultant in Figure 3.3a. This simple graph demonstrates that complex vibration data contains multiple frequencies, which all contributing various amplitudes to the total vibration measured by vibration transducers, but then how may we graphically depict these Fourier frequency and amplitude components? First, we use modern computer technology by performing a mathematical conversion from the "time to frequency domains" and next plot the result graphically. Figure 3.4a is identical to Figure 3.3b. The Fourier transformation from the time to frequency demains for this waveforms is shown in Figure 3.4b. Here the horizontal axis represents

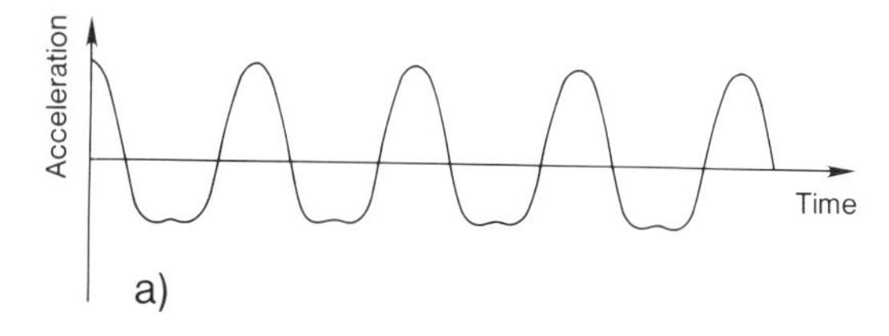

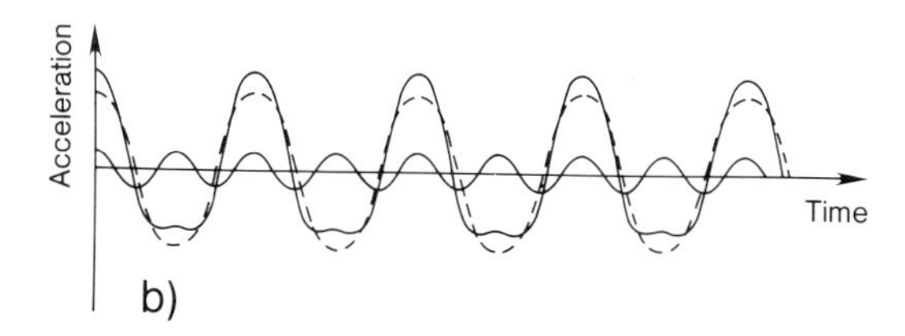

Fig. 3.3 (a) Periodic vibration containing two frequencies, (b) same vibration as above but showing each of the two vibration frequencies and their summation waveform. (Courtesy B & K Co.)

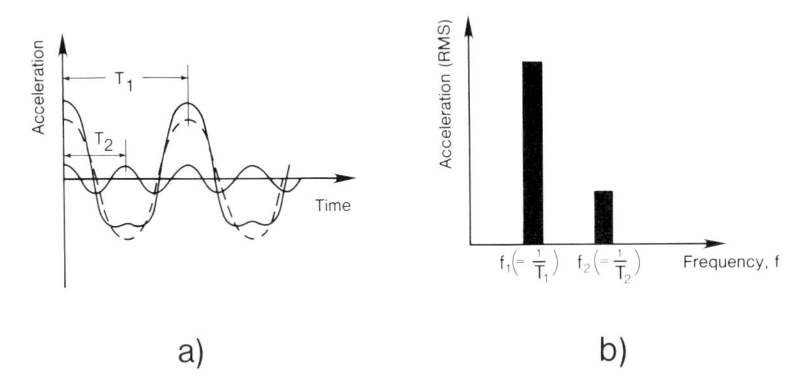

Fig. 3.4 (a) Same vibration as in Fig. 3.3(b) in the time domain converted into its equivalent frequency domain spectrum in 3.4(b). (Courtesy B & K Co.)

frequency, the vertical axis represents magnitude (e.g. acceleration.) The number of vertical lines indicates the total number of vibration frequencies present in the Fourier spectrum; the horizontal position of each line defines a single vibration frequency; the height of each (frequency) line determines the vibration acceleration amplitude which that frequency contributes to the

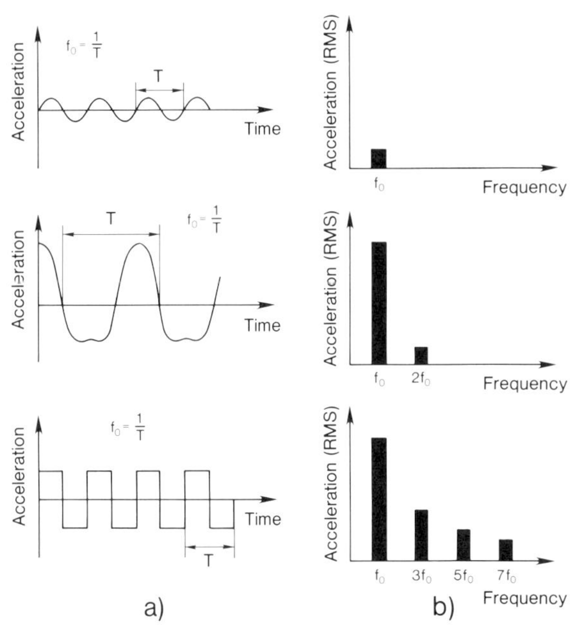

Fig. 3.5 (a) Various vibration acceleration waveforms converted into their respective equivalent frequency domain spectra, in 3.5(b). (Courtesy B & K Co.)

total spectrum. Thus the displayed Fourier spectrum of Figure 3.4a is shown to have two frequency components (f1 and f2). Their relative vertical (bar) heights correspond to their relative amplitude contributions to the overall spectrum. Since most vibration in the workplace is unique, Fourier analysis can be used to characterize, or "finger print" each vibration environment. Furthermore, vibration spectrum is usually different depending on the direction (axis) of the vibration measurement Figure 3.5 further illustrates this point. In Figure 3.5 a we are given our familiar periodic sinusoid; and similarly a complex periodic square wave function. Their corresponding Fourier Spectrums are shown in Figure 3.5b. It can be seen that the square wave spectrum contains a series of four peaks containing the fundamental frequency and its harmonics.

The advent of high speed digital computers has made it possible to produce spectrum in real time. Special computer algorithms (called Fast Fourier Transforms-FFT) have been devised to speed the computer processing. All of this has made spectrum analysis much easier to perform directly in the workplace.

3.4 RANDOM VIBRATION (3)

Random variation may be defined as a continuing oscillating motion in which the acceleration varies in a non-periodic manner with time. The waveform does not repeat itself in a cyclic manner but is irregular with time. Random vibration because of its non-periodic nature (as in the case of a mechanical shock pulse) results in a continuous frequency spectrum.

An example of random motion is given in Figure 3.6, in which instantaneous acceleration is plotted against time. Both the magnitude of the peaks and the period of time between zero acceleration vary irregularly. A motion of this sort is the result of a very large number of events occurring by chance. The total acceleration waveform can be considered composed of many sinusoids, each moving at its own frequency with different magnitudes that vary in time.

When it is required to predict with some accuracy the probabilities of the occurrences of random events, a suitable probability law must be employed. From the nature of the continuous frequency spectrum of a random vibration it is reasonable to assume the instantaneous accelerations in a given frequency band have a "normal probability distribution" about zero as a mean. A plot of the relative frequency of occurrence or "probability density" of an acceleration which follows the normal probability law is shown in Figure 3.7 where $a/\bar{a}$ is the ratio of the instantaneous acceleration "a" to the standard (rms) acceleration "$\bar{a}$".

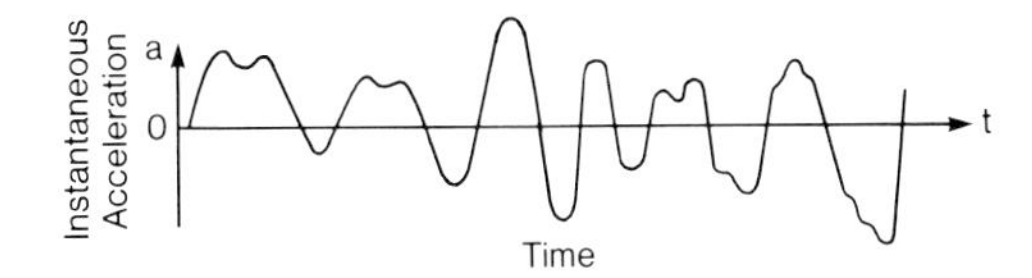

Fig. 3.6 Acceleration vs. time waveform of typical random motion. (Courtesy Columbia Research Co.)

Figure 3.7 is useful since the area under the curve P_m between any two limits is the probability that a typical value of a/a will fall within these limits. Thus, in the case of $a/\bar{a} = +1$ we can say the instantaneous acceleration "a" is equal to or less than the standard deviation 68.3% of the time.

A quantity which gives a concise picture of the spectral distribution when plotted as a function of frequency is the so-called "acceleration density" and is given by:

$$G(f) = \lim_{B \to 0} \frac{\bar{a}^2}{B} \tag{24}$$

where, G(f) = the mean acceleration density

a = the root mean square of the random accelerations

B = the bandwidth or frequency range from f_1 to f_2 under consideration.

The mean acceleration density G(f) is in units of g^2/Hz. Thus, the acceleration density is often referred to under various names as the mean square acceleration, spectral density, mean square acceleration per cycle, or the power density.

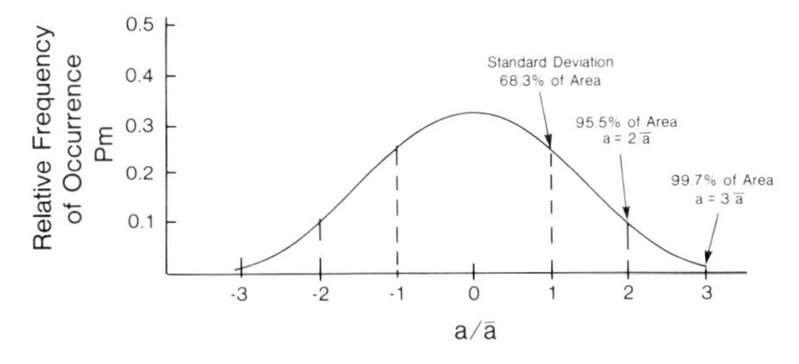

Fig. 3.7 Gaussian or normal distribution curve. (Courtesy Columbia Research Co.)

Since the acceleration density is a function of frequency, the rms acceleration in the band between f_1 and f_2 can be calculated from equation (25):

$$\bar{a}^2 = \int_{f2}^{f1} G(t)dt \qquad (25)$$

Simply stated, equation (25) defines the rms acceleration as the square root of the area under curve in Figure 3.8 between frequencies f_1 and f_2 and is indicated, by the shaded area. As the bandwidth is increased the observed rms acceleration $\bar{a}$ will continue to increase until the entire area below the acceleration density curve G(f) is included.

A random vibration having a constant acceleration density with frequency is referred to as "white" and contains a uniform or flat frequency spectrum. In this case the acceleration density becomes independent of bandwidth and equation (24) becomes:

$$G_0 = \frac{\bar{a}}{B} \qquad (26)$$

where: G_0 = a constant or white acceleration density.
$\bar{a}$ = acceleration in $g^2{}_{rms}$ units

As an example: Find the root mean acceleration a from an actual test calling for a constant acceleration $0.2g^2/H_z$ over the frequency bandwidth B from 15 to 2015 Hz then solving for a:

$$a = G_0 B \qquad (27)$$

Substituting into equation (27): $G_0 = 0.2g^2/\text{Hz}$, B = (2015 - 15) Hz

$a = (2015 - 15)(0.2)$

$a = 20\ g_{rms}$

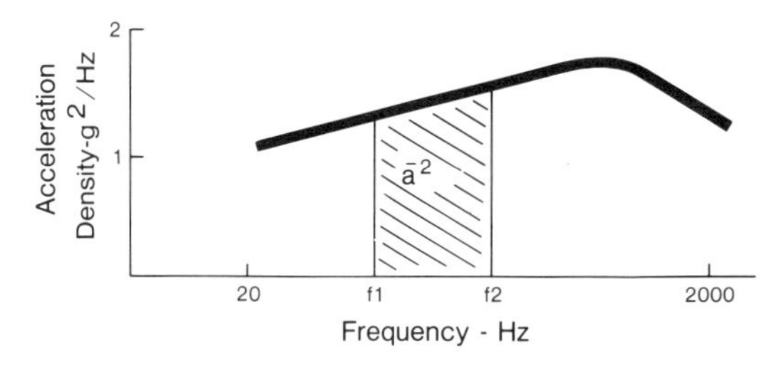

Fig. 3.8 Typical plot of acceleration density as a function of frequency. (Courtesy Columbia Research Co.)

Here, with the aid of the normal distribution law plotted in Fig. 3.8 we find that the instantaneous acceleration $\bar{a}$ = 20 $g^2{}_{rms}$ occurs 69% of the time. The magnitude of the instantaneous acceleration "a" will exceed twice the rms value 40 g_{rms} about 5% of the time and will be three-fold the rms value to 60 g_{rms} 0.3% of the time.

3.5 IMPULSE FUNCTIONS

Sometimes in vibration work impulse types functions occur. It is beyond the scope of this text to discuss these except to state that it is possible with difficulty to perform a Fourier spectrum analysis of these functions, and examine their frequency and amplitude content as with vibration. Common examples are given in Fig. 3.9.

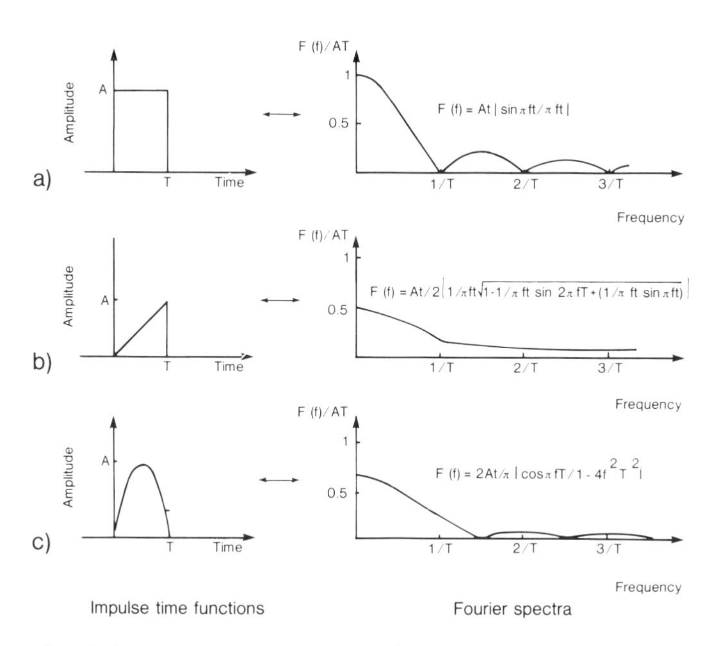

Fig. 3.9 Examples of shock time pulse functions (a,b,c) and their respective Fourier spectra.
(Courtesy B & K Co.)

3.6 OTHER VIBRATION TERMS

3.6.1 Vibration Transmissibility

Is defined for each vibration frequency present in the spectra and, is the ratio of vibration acceleration appearing at one point divided by impinging vibration at another point, where both are applied in the same direction.

Transmissibility is a method of determining how vibration is modified as it moves through a structure (in our case the human body): a ratio of one means there is no change in the vibration applied at one point and appearing at a second point; a ratio greater than one indicates an amplification of the original vibration; a ratio less than one indicates attenuation of the original vibration.

3.6.2 Coherence

Is a measure of the "goodness" of the vibration transmissibility data. The vibration input to a structure and the responding output are both assumed to be linear. A coherence of "1" indicates that the measured vibration appearing at a second point, thought to be due to vibration impinging upon the first point, is truly derived from this first point and not from any other vibrating source. A coherence of "0" indicates that the measured vibration appearing at a second point is not due solely to the vibration applied to the first point. A value between 0 and 1 is suspect, with possible speculation that the structure under examination is nonlinear.

Mathematically, if we assume a linear system, the coherence function is defined as:

$$\gamma^2 = \frac{Gyx^2}{(Gxx)(Gyy)} \qquad (28)$$

where: Gyx = cross power spectrum between input and output
Gxx = power spectrum of input
Gyy = power spectrum of output

3.6.3 Resonance

Is the tendency of the human body (or any other mechanical system) to act in concert with externally generated vibration and to actually amplify the impinging vibration. For vertical vibration, in the region of 4 to 8 Hz, more particularly 5 Hz, seated man's "whole-body" (i.e., mostly upper torso) is in resonance and thus there is an amplification of impinging vibration in various amounts depending on where measured. Furthermore, various body parts resonate at other frequencies: for example the head-shoulder system resonate in the 20 to 30 Hz range, the eyeball resonates in the 60 to 80 Hz range, to name a few. In general the larger the system mass, the lower the resonant frequency. The concept of resonance is a crucial one since it would appear that at resonance frequencies man is most likely to be most susceptible to the effects of vibration exposure than at other frequencies.

3.6.4 Mechanical Impedance

(Z) is the ratio of applied force to resulting velocity. It is called driving point impedance if both the force and velocity are measured at the same point. One can apply force at one point and measure the resulting velocity at another point, this is called transfer impedance.

$$Z = \frac{\text{Force}}{\text{Velocity}} \tag{29}$$

3.6.5 Mechanical Mobility

Mechanical mobility is mechanical admittance and is defined as the inverse of mechanical impedance.

3.6.6 Mechanical Stiffness

Mechanical stiffness is the ratio of the change in force to the corresponding change is displacement of an elastic element:

$$\text{Stiffness} = \frac{\text{Force}}{\text{Displacement}} \tag{30}$$

3.6.7 Mechanical Compliance

Mechanical compliance is the ease with which a system may be displaced or compressed and is the reciprocal of stiffness.

3.6.8 Dynamic Mass

Dynamic mass is defined as the ratio of applied force to its resulting acceleration:

$$\text{Dynamic Mass} = \frac{\text{Force}}{\text{Acceleration}} \tag{31}$$

3.6.9 Dynamic Modulus

Dynamic modulus is the ratio of stress to strain under vibratory conditions:

$$\text{Dynamic Modulus} = \frac{\text{Stress}}{\text{Strain}} \tag{32}$$

3.7 DYNAMIC MODELING OF MECHANICAL AND HUMAN SYSTEMS

Engineers traditionally tend to "model" or characterize a system in an effort to understand how it operates and how the system changes when operating parameters are varied. In this section a brief discussion of mechanical modeling and how it applies to the human system under vibration is presented. (The reader is referred to other references (1,2,4-6) and the supplementary bibliography on human modeling and mechanical modeling, for an indepth treatment of the subject.)

Three idealized passive elements form a vibration system, they are a mass (m), spring (k) and a damper (or dash pot) (c). In Fig. 3.10, there is an excitation force F(t) applied in the direction shown moving a simple single rigid mass system away from its resting equilibrium position. In order to describe this motion, equations of motion are next developed based upon Newton's Second Law of motion: Force = mass x acceleration (equation 33). The resulting motion occurs in the direction in which the force acts on the system:

$$F = ma = m\frac{d^2x}{dt^2} \tag{33}$$

The spring motion is best described by compression or expansion of the spring itself (considered to have neglible mass) determined by Hooke's law, namely, the spring force is proportional to the spring's deformation or a constant of proportionality "K" time the actual distance the spring moves "x":

$$F = Kx \tag{34}$$

The motion of the damper (or dashpot) is proportional to a (coulomb) damping force "c" times the speed or velocity of the motion

$$F = cV = c\frac{dx}{dt} \tag{35}$$

A positive and negative direction of motion is chosen and the following equation is obtained first <u>without</u> an external vibration force (equation 36) and then motion <u>with</u> a vibration force f(t) (equation 37) for the system as a whole:

$$F = ma + cV + Kx = m\frac{d^2x}{dt^2} + c\frac{dx}{dt} + Kx = 0 \tag{36}$$

$$F = m\frac{d^2x}{dt^2} + c\frac{dx}{dt} + Kx = f(t) \tag{37}$$

The forms of equations (36) and (37) are known as "differential equations of motion" and must be solved by the methods of advanced calculus (beyond the scope of this book). The system described is a linear system and is called a "single" degree-of-freedom system since the model consists of one mass moving along one axis. Part of the mathematical solution to equation (37) involves determining at what frequencies the system will "resonate" and what the effects on the system are by damping (i.e., means of dissipating vibration energy within a vibrating system). In the case of a single degree of freedom

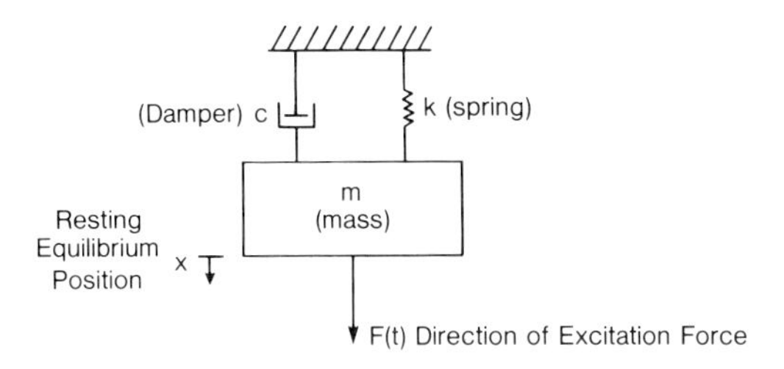

Fig. 3.10 Basic elements of a single-degree-of-freedom mechanical vibration system.

system as described above, resonance W_o turns out to be equation (38)

$$W_o = \left(\frac{K}{m}\right)^{1/2} \tag{38}$$

where $W_o = 2\pi f_o$.
Solving for the resonant frequency f_o we get:

$$f_o = \frac{1}{2\pi}\left(\frac{K}{m}\right)^{1/2} \tag{39}$$

Thus in this case the resonance frequency of the system depends only on the values spring constant K and the system mass m.

The effects of damping for this system is given by equation (40) (see fig. 3.11)

$$Q = \left(\frac{Km}{c}\right)^{1/2} \tag{40}$$

Where Q is called the "quality factor." The larger the Q value the smaller damping. The smaller the Q value the larger the damping. When Q is infinitely large the system is totally undamped. When Q = 1/2 the system is called "critically damped." A critically damped system is usually desirable since the system response to vibration is optimum. An over damped system is sluggish and does not respond well to incoming vibration. An underdamped system tends to ring or oscillate and overreacts to incoming vibration. The effects of system damping and system response for our example is shown in Figure 3.11 a and b. Figure 3.11a shows that for less damping (larger Q values) the system amplitude response to excitation grows larger and larger, whereas at lower Q values (e.g. critical damping) the system calms down quickly and does not produce high amplitude responses. The corresponding phase angle between system response and excitation is given for the various Q values in fig. 3.11b. A number of interesting facts can be seen from these

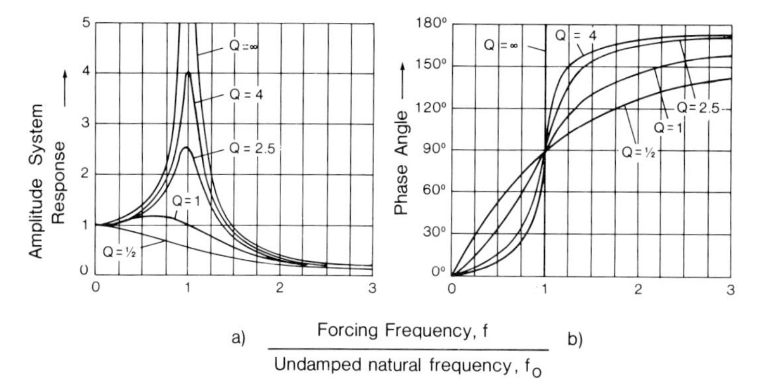

Fig. 3.11 Examples of complex frequency response functions showing absolute values of the system response in (a) with the respective phase lag between response and excitation in (b).
(Courtesy B & K Co.)

phase curves (1,2):

1.) In the case of no damping (Q = ∞) the response and the excitation are in phase (0 = 0) below resonance, while above resonance they are 180° out of phase. Because Q = ∞ the change in phase takes place in the form of a discontinuous jump.

2.) When Q = ∞, i.e. damping is introduced in the system, the change in phase between response and excitation tends to take place gradually, and the larger the damping (the smaller Q) the slower is the phase change with frequency around resonance.

3.) Independent of the magnitude of the damping, the phase lag between the response and the excitation at resonance is 90°.

If the system being studied consists of several masses interconnected with springs and damper elements the measure of Q stated above cannot be utilized unless the coupling between the different masses is so small that a unidirectional motion of one mass does not influence the motion of any of the others (or vice versa).

Systems in which a single mass moves in more than one direction or systems which consist of several, elastically interconnected masses, are commonly called multi-degree-of-freedom systems. A linear multi-degree-of- freedom system can be mathematically described by a set of linear differential equations of motion similar in form to equations 36 and 37. When the frequency response curve of the system is plotted it will normally show one resonance "peak" for each degree-of-freedom. Thus, a two degree-of-freedom system shows two resonance peaks, a three degree-of-freedom system shows three resonance peaks, etc., (see Figure 3.12).

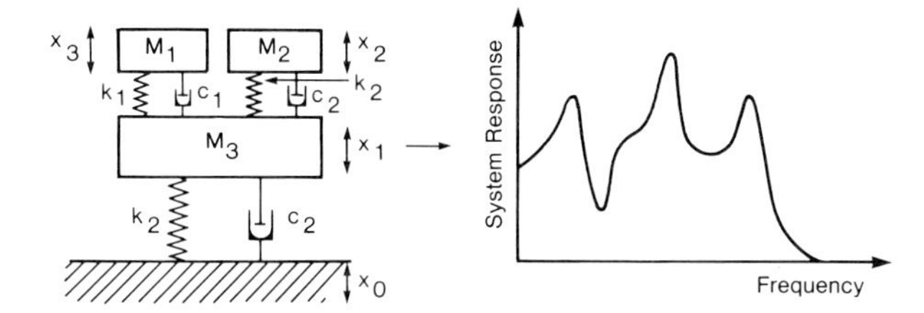

Fig. 3.12 Example of a three-degree-of-freedom mechanical vibration system with its corresponding frequency response function. (Courtesy B & K Co.)

In summary, at system resonance(s) unless there is sufficient damping, the system overresponds producing an output response far in excess of the excitation input, with too much damping, the system is sluggish and underreacts. Although we have briefly described a relatively simple one-degree-of-freedom system, much of these same principles apply to human resonances as well.

The resonance concept applies directly to reducing or mechanically isolating sources of vibration in the workplace. For example, Figure 3.13 is a typical machine's frequency response plot of vibration transmissibility, the ratio of the vibrating excitation forcing frequency "f" to the undamped natural frequency "f_o." As it turns out in order to produce isolation we want the isolators own resonant frequency different from the machines resonant frequency, in this case one selects isolators with f_o such that the lowest machine forcing frequency is greater than 1.4 f_o. The trade-off is a high damping ratio protects best against vibration transmission, as the vibrating machine passes through the isolators resonance and it builds up operating speed. However, a low damping ratio gives the best protection at high forcing frequencies.

Modal Analysis (7): Most mechanical systems under vibration require a large number of differential equations of motion to be solved simultaneously in order to characterize the effects of vibration impinging on the total system. With the advent of integrated electronic circuits and modern, high speed, portable computers has emerged a powerful new software analysis technology called "modal analysis." Modal analysis is a way of describing the response of an object to vibration forces that act upon it. For example, consider the response of a bell when struck: it rings. The precise manner in which the bell rings depends upon many factors. Using the principle of modal analysis, the ringing can be broken down to a summation of a number of elements called "normal mode shapes" which when summed produce the original sound and motion of the bell. Each of the modes is characterized by unique

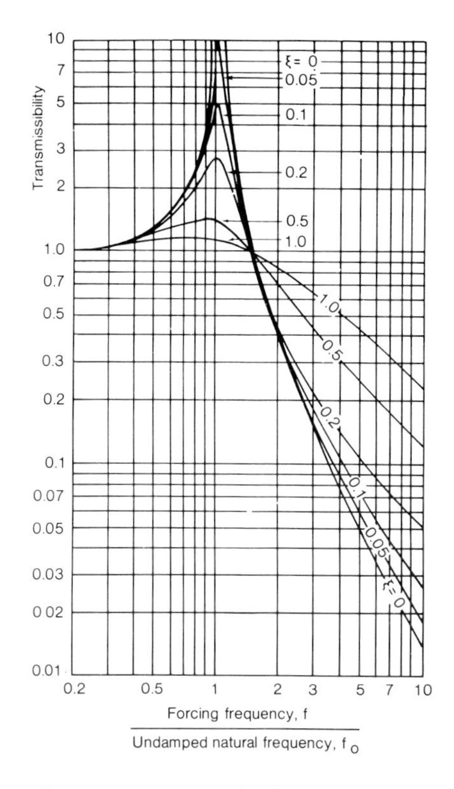

Fig. 3.13 Transmissibility curves for isolation selection. (Courtesy B & K Co.)

modal parameters: resonant frequency, damping, and mode shape. For the bell, each natural or resonant frequency corresponds to a musical tone it produces, damping governs the duration of the tone, while the mode shape describes the pattern of vibration. The overall timber of the bell is determined by the combination of all modes present. Thus the bell sounds a little different depending upon how and where it is struck since the relative strength of each mode is governed by the exciting force. Despite the simplicity of this example, the idea of modal analysis in fact applies to many moving mechanical systems (Figure 3.14).

Modal analysis can be performed either experimentally or analytically. In both cases the goal is to determine the modal parameters, and thus develop the modal model that describes the dynamic properties of the object under study.

The modal model is actually a mathematical model of the test object. It is based on the object's modal parameters: resonant frequencies, modal damping, and mode shapes. Traditionally, the model was a reduced set of the differential equations of motion, written in terms of the modal parameters.

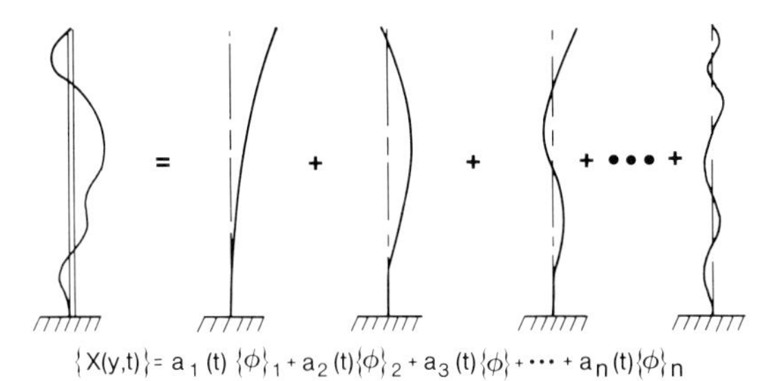

Fig. 3.14 The modal analysis concept.
(Courtesy B & K Co.)

The modal parameters themselves are determined by the physical characteristics of the vibrating object and its components: mass, stiffness, and damping. In this new analytical approach of modal analysis, these physical characteristics are modeled directly using mass, stiffness, and damping elements.

In modal testing there is no need to measure the physical characteristics of the test object explicitly. A set of frequency response functions is measured and provides all of the information necessary to solve the differential equations of motion.

Modal testing can be performed in a number of ways. All methods involve measuring the relation between a force input (excitation) and a motion output (response) between various points on the structure. In the most common method, the force input is applied at a single point, and the relation between input and output is measured as a function of frequency. This approach is called single-point excitation.

The technique lends itself to rapid measurements using a dual-channel FFT spectrum analyzer, and subsequent data analysis using a small, transportable desktop computer. The excitation force may be provided by a vibration shaker system, or other means.

As an example, Figure 3.15 shows sample results for one vibration mode of a plate-like structure. Vibration Excitation was provided to the structure, and 25 frequency response measurements were made for the structure using a dual-channel spectrum analyzer. In this figure we see the computer graphic of first the plate-like structure with no vibration (in three coordinates), and the resulting graphic of the vibration mode shapes produced by vibration excitation. Finally, the limitation this analysis assumes that the system under vibration produces a <u>linear</u> output response to an excitation input, however, as we explore the human response to vibration we find that the human response to vibration is not necessarily linear.

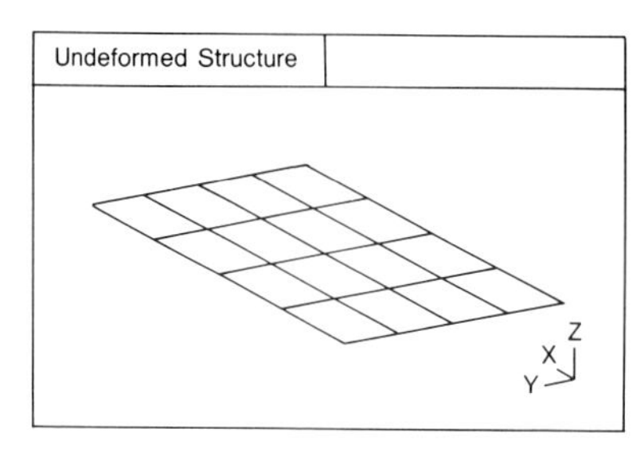

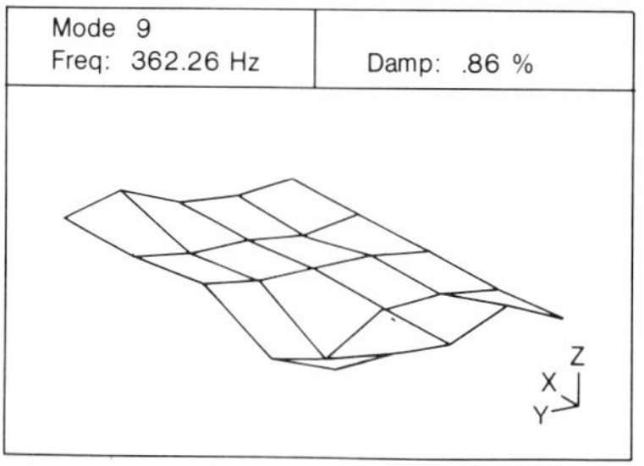

Fig. 3.15 Example of modal analysis used for a plate-like structure. (Courtesy B & K Co.)

Human Biodynamic Models: Through the years several vibration researchers have attempted to model human response to vibration (and impact) in the laboratory. Whole-body modelling originally began with Von Bekesy (8) and continued later with the simultaneous definitive experiments of the late Rolf Coermann in the U.S. and the Dieter Dieckmann in Germany in the early 1960's in an effort to discover the fundamental human whole-body resonances on which much of our knowledge is built on today.

Coermann's studies (5,9,10) of human mechanical impedance showed that for seated man whole-body resonance to vertical vibration appears between 4-8 Hz (nominally 5Hz) (see Figure 3.16 a & B). When subjects are standing not only does the characteristic 5 Hz resonance appear, but also a second less prominent resonance appears from 10-14 Hz (nominally 12 Hz). These resonances varied somewhat depending on whether the subject was sitting or standing erect or relaxed (see Figure 3.17). Coermann compared his results to a pure mass, pure spring, pure damper, and combination systems (see Figures 3.16, 3.17, 3.18) in an effort to mathematically model the whole-body system under vibration, resulting in the mechanical model shown in Figure 3.19. Independently and simultaneously Dieckmann (11) performed human transmissibility studies (see Figure 3.20) arriving at virtually the same conclusion as Coermann that (mostly upper torso) whole body resonance occurs in the approximately 4-8 Hz range. Figure 3.20 also shows head resonances in

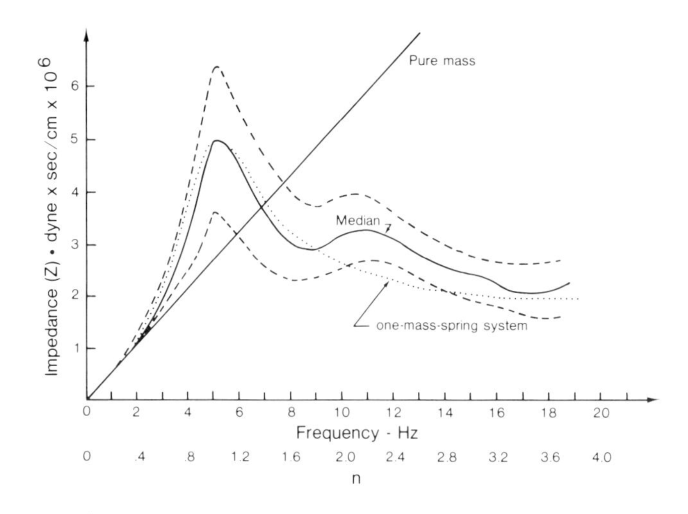

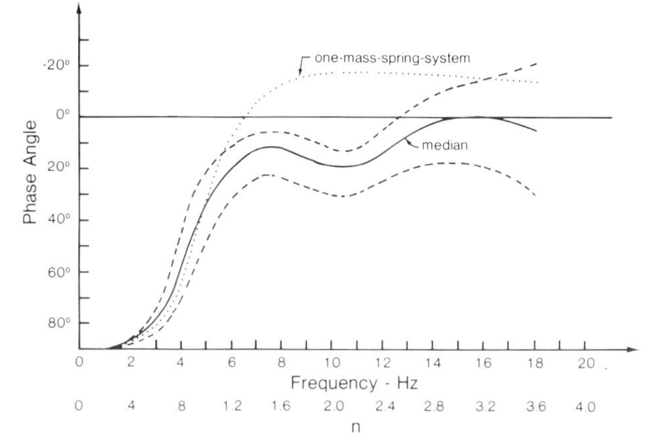

Fig. 3.16 (a) The median and modulus of mechanical impedance for 8 human sitting subjects, compared with the impedance of a pure mass and a one-mass-spring system with 0.636 damping coefficient. (b) corresponding phase angles for (a) above.
(Adapted from Coermann, reference (5))

the 30Hz range. In the late 1960's Miwa in Japan published an extensive series of studies (12-17) based upon numerous parameters and conditions under whole-body vibration. In many ways Miwa's work substantiated and enhanced the early work of Coermann and Dieckmann.

In the hand arm vibration area numerous models have been proposed by Dieckmann (18) Reynolds (19) (see Figure 3.21), Suggs (20,21), Abrams (22), Mishoe (23), and Miwa (24). Although each of these models has merit, unfortunately, none of them have successfully included the interaction of hand-arm vibration, the model, and vibration white finger (VWF) resulting from

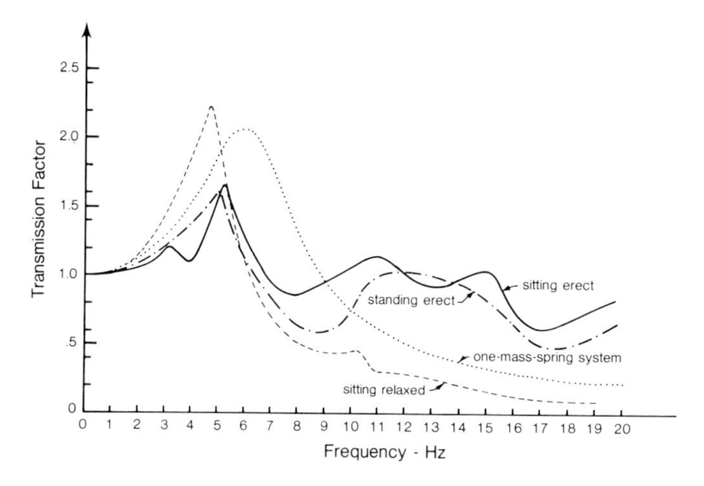

Fig. 3.17 The vibration transmissibility from seat to the head of one subject at varied body postures compared to the transmissibility of a one-mass-spring system with damping. (Adapted from Coermann, reference (5))

the impedance of a pure mass is

$$\bar{Z} = j m \omega$$

of a pure spring

$$\bar{Z} = -j \frac{k}{\omega}$$

of a pure damper

$$\bar{Z} = c$$

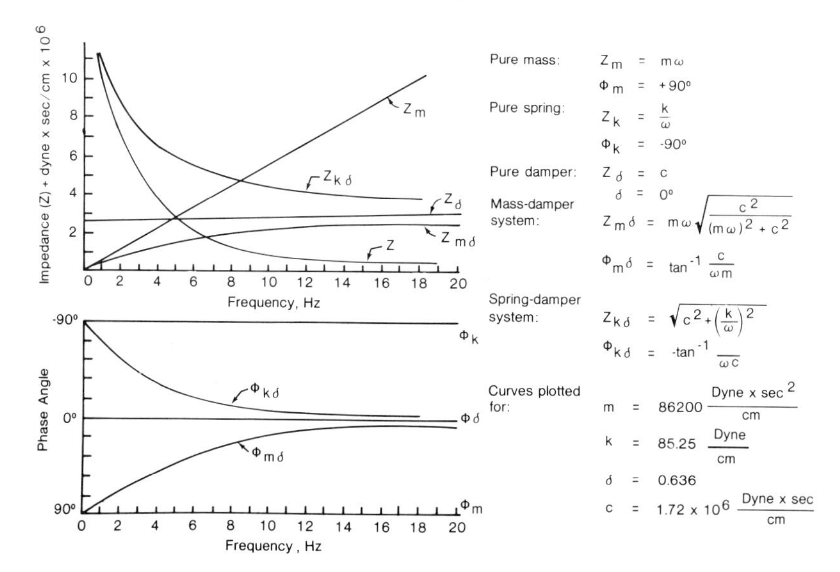

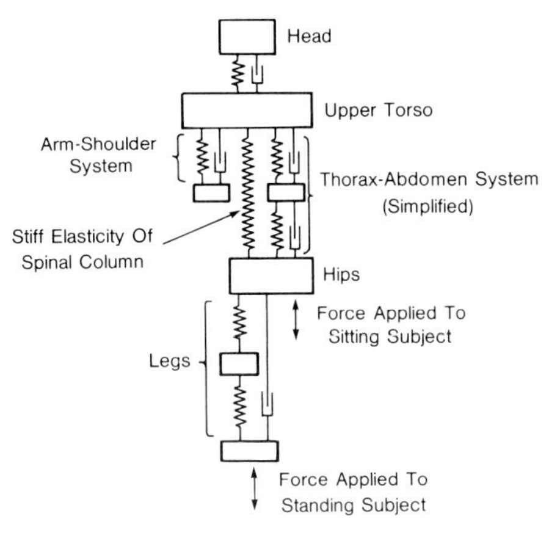

Fig. 3.19 Simplified model of whole-body mechanical system of the human body standing on a vertical vibration shaker system.
(Adapted from Coermann, reference (10))

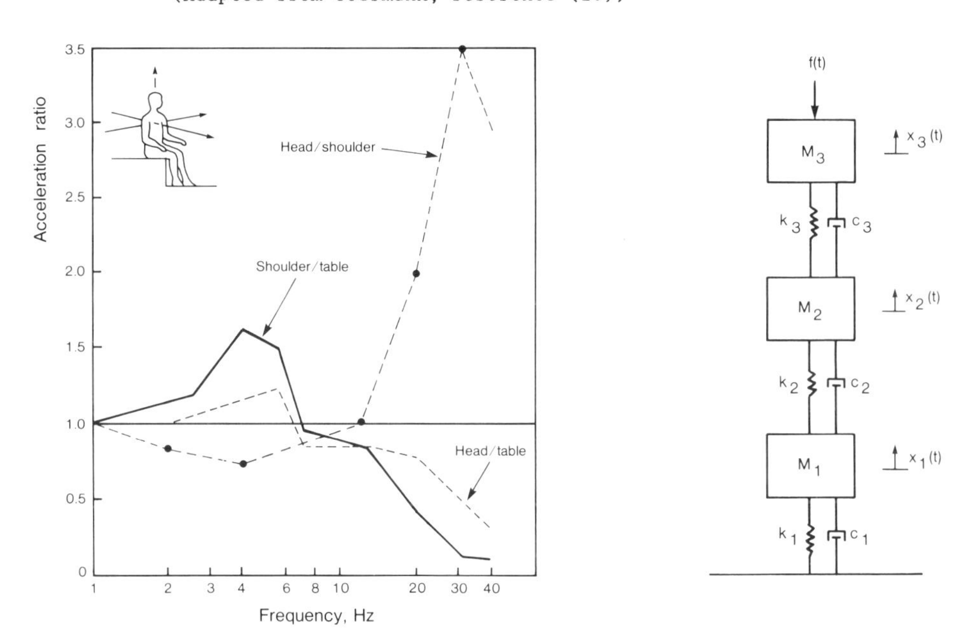

Fig. 3.20 Transmissibility of vertical vibration from vibrating table to various body parts for a seated human.
(Adapted from Dieckmann, reference (11))

Fig. 3.21 Model of a three-degree-of-freedom mass, spring, damper hand-arm system.
(From Reynolds, reference (19))

Fig. 3.18 Impedance and phase characteristics of pure mass, pure spring, pure damper, mass-damper, and spring damper systems.
(Adapted from Coermann, reference (5))

exposure. Unfortunately, inconsistencies in and between the various models exist because different assumptions and data were used to derive each of the models. Thus, there is not a clear picture of hand-arm vibration parameters (eg resonances) as there is with whole-body resonance.

Finally, as vibration in the workplace is explored in the next chapter, the physical concepts presented in this chapter will aid in an understanding as to how the worker reacts to vibration and how to measure the vibration impinging on him.

REFERENCES

1 J.T. Broch, Mechanical Vibration and Shock Measurements, Bruel and Kjaer Co., Naerum, Denmark, 1973.
2 R.B. Randall, Frequency Analysis, Bruel and Kjaer, Co., Naerum, Denmark, 1977.
3 Technical Staff, Shock and Vibration Handbook, Columbia Research Labs., 1971.
4 F.S. Tse, I.E. Morse, and R.T. Hinkle, Mechanical Vibrations, Allyn and Bacon Publ., Boston, 1971.
5 R.R. Coermann, The Mechanical Impedance of the Human Body in Sitting and Standing Position at Low Frequencies, U.S. Air Force Aerospace Medical Reseach Lab. Report No. 61-492, 1961. (Ibid. Human Factors, 4 (1962) 225-253.
6 Proceedings of the Symposium of Biodynamic Models and Their Applications, Bergamo Center, Dayton, Ohio, February 15-17, 1977.
7 Technical Staff - Application Brief: Modal Analysis-An Introduction, Bruel and Kjaer, Co. Naerum, Denmark, 1984.
8 G. von Bekesy, Ueber die Empfindlichkeit des Stehenden und Sitzenden Menschen gagen sinusfoermige Erschuetterungen, Akustische Zeitschrift, 4 (1939) 360-369.
9 R.R. Coermann, Comparison of the Dynamic Characteristics of Dummies, Animal, and Man, U.S. National Academy of Science National Research Council Publication 977, 1962.
10 R.R. Coermann, The Passive Mechanical Properties of the Human Thorax-Abdomen System and of the Whole-Body System, J. Aerospace Medicine, 31 (1960) 915-924.
11 D. Dieckmann, Einfluss vertikaler mechanischer Schwingungen auf de Menschen, Internationale Zeitschrift fuer Angewandte Physiologie Einschliesslich Arbeitsphysiologic, 16 (1957) 519-564.
12 T. Miwa, Mechanical Impedance of the Human Body in Various Postures, Ind. Health (Japan), 13 (1975) 1-22.
13 T. Miwa, Evaluational Methods for Vibration Effects - Response to Sinusoidal Vibration At Lying Posture, Ind. Health (Japan), 7 (1969) 116-126.
14 T. Miwa, Evaluation of Methods for Vibration Effects - The Vibration Greatness of the Pulses, Ind. Health (Japan), 6 (1968) 143-164.
15 T. Miwa, Evaluation of Methods for Vibration Effects - Measurement of Threshold and Equal Sensation Contours of Whole-Body for Vertical and Horizontal Vibrations, Ind. Health (Japan), 5 (1967) 103-205.
16 T. Miwa, Evaluation of Methods for Vibration Effects - Measurement of Equal Sensation level for Whole-Body Vibrations Between Vertical & Horizontal Sinusoidal Vibration, Ind. Health (Japan), 5 (1967) 206-212
17 T. Miwa, Evaluation of Methods for Vibration Effects-Calculation Methods of Vibration Greatness level on Compound Vibrations, Ind. Health (Japan), 6 (1968) 11-17

18 D. Dieckmann, Ein Mechanisches Modell fur das Schwengungserregte Hand-Arm System des menschen, Internatl. Z. Angew. Physiol. Einschl. Arbeisphysuik, 17 (1958) 25.
19 D. Reynolds, Hand-Arm vibration: A Review of Three Year's Research, Proc. of the Internatl. Occup. Hand-Arm Vib. Conf. (D. Wasserman and W. Taylor Eds.), 99-128, DHEW/NIOSH Public No. 77-170, 1977.
20 C.W. Suggs and J.W. Mishoe, Hand-Arm Vibration: Implications Drawn from Lumped Parameters Models, (Ibid), 136-141.
21 C.W. Suggs and L.A. Wood, A Distributed Parameter Dynamic Model of the Human Forearm, (Ibid) 142-145.
22 C.F. Abrams, Modeling the Vibration Characteristics of the Human Hand by the Driving Point Mechanical Impedance Methods, Unpublished Ph.D. dissertation, Dept. of Biol. & Agric. Engr., North Carolina State Univ. at Raleigh, 1971.
23 J.W. Mishoe and C.W. Suggs, Hand-Arm Vibration Part 1, Subjective Response to Single and Multi-Direction Sinusoidal and Nonsinusoidal Excitation, J. Sound and Vibration, 35(4) (1974) 479-488.
24 T. Miwa, Evaluation Methods for Vibration Effects - Measurement of Threshold and Equal Sensation Contours on the Hands for Vertical and Sinusoidal Vibrations, Ind. Health (Japan), 5 (1967) 213-220.

Supplementary Bibliography

J.S. Bendat, Principles and Applications of Random Noise Theory, John Wiley, & Sons, Publishers, New York, 1958.

J.S. Bendat and A.G. Piersol, Measurement & Analysis of Random Data, John Wiley and Sons, Publishers, New York, 1965.

S.M. Crandall and W.D. Mark, Random Vibrations in Mechanical Systems, Academic Press, New York, 1963.

C.T. Morrow, Shock and Vibration Engineering, John Wiley and Sons, Publishers, New York, 1963.

International Standards Organization, #2041 - Vibration and Shock Vocabulary, Geneva, 1975.

H.E. Von Gierke, Biodynamics Response of the Human Body, Applied Mechanics Reviews, 16 (1964) 951-958.

H.E. Von Gierke and R.R. Coermann, The Biodynamics of Human Response to Vibration and Impact, Ind. Med. and Surgery, 32 (1963) 30-33.

Proceedings at the U.K. Informal Group on Human Response to Vibration, Institute of Sound and Vibration Research, Univ. of Southampton, U.K., Sept. 1975.

Chapter 4

Instrumentation and the Measurement and Evaluation of Vibration in the Workplace

4.1 INTRODUCTION

In this chapter the important elements of instrumentation for and performing vibration measurements and evaluations in the workplace are discussed. Also, discussed are how to choose an appropriate measuring transducer, methods of calibrating transducers, typical signal conditioning configurations, and the basics of spectrum analyzers.

4.2 MEASUREMENT COORDINATE SYSTEMS

Since vibration is a vector quantity, magnitude and direction and placement of the measuring transducer must be specified. Both hand-arm vibration and whole-body vibration measurements are obtained with respect to internationally agreed upon biodynamic coordinate systems. (See Fig. 4.1.) International Standards Organization (ISO) #5349 (1) specifies the hand-arm coordinate system used which is defined at the third metacarpal of the hand. However, it is not always possible to mount a transducer directly at this point on the hand, and more often than not the transducer is mounted on the tool handle per se, where the hand grips the tool handle (basicentric system, Fig. 4.2). The terms Z_h, X_h, Y_h define the three orthogonal directions used for measurements.

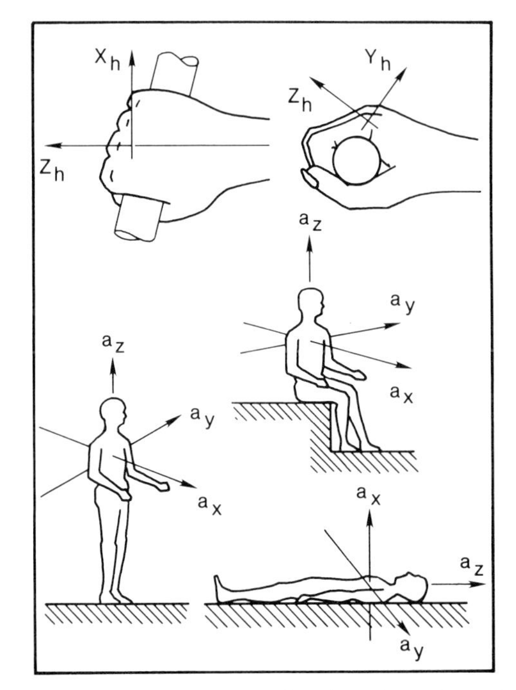

Fig. 4.1 ISO 5349 (hand-arm) and ISO 2631 (whole-body) biodynamic coordinate systems for human measurements. (Courtesy B & K Co.)

Similarly ISO #2631 (2) specifies the biodynamic coordinates for performing whole-body measurements and defines the three planes of motion impinging on the body. (In these cases rotational motion is usually not considered, only linear motion is defined.)

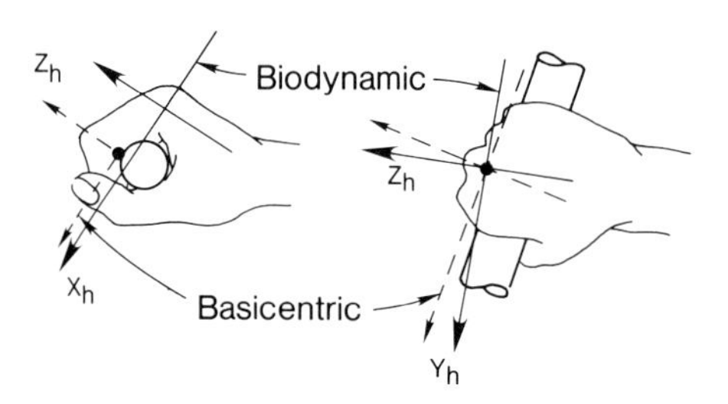

Fig. 4.2 Biodynamic and basicentric coordinate systems for hand-arm and tool measurements.
(Courtesy ACGIH)

4.3 VIBRATION TRANSDUCERS

In many countries including the U.S. the measurement of choice is vibration acceleration, and not necessarily velocity or displacement in human vibration measurements, for these reasons: 1) There are numerous accelerometer types and configurations available, 2) Obtaining an acceleration measurements allows one to easily obtain corresponding velocity and displacement information using electronic integrators. Thus from one measurement, three parameters are obtained, 3) It is believed by some researchers that it is acceleration which produces potential harm to the body (3-6) and thus a direct measurement of acceleration is desirable.

4.4 ACCELEROMETERS

There are basically three types of accelerometers used in human vibration work: piezoelectric (or crystal accelerometers), piezoresistive, and strain gage. Each type of device is unique and has various measurement advantages and disadvantages depending on its application. Before discussing each device in some detail it is well to discuss some of the criteria used in choosing the correct device for the desired measurement. First, it is the work situation, or that which is to be measured, which ultimately determines which device is suitable for measurements. In particular, the following parameters must be considered: a.) The <u>mass</u> of the accelerometer MUST be as small as possible in relation to the structure to be measured. In order to avoid inaccuracies due to a large mass devices (called "mass loading"), the rule of thumb is that the accelerometer mass should be no more than one tenth of the dynamic mass of the vibrating point onto which it is mounted (see Fig 4.3). For human measurements mounted to the body 1-2 gram accelerometers (15 grams maximum) are desirable. On the other hand, accelerometers measuring, for example, vehicular truck dynamics on the cab floor could weigh several grams and still obtain accurate measurements (see Fig. 4.2). b) Dynamic amplitude range of the device must be such as to accommodate the maximum acceleration level anticipated. For example, it would be folly to use a 10 g accelerometer to measure a 1000 g acceleration source. The device would be destroyed very quickly. Similarly, it would be folly to use a 1000 g accelerometer to measure 1 g signal because the device would be insensitive to small signals and small changes would never be seen. It is true, that with today's modern electronics technology, low noise, high gain preamplifiers could be used to amplify this small signal, but why go to all the trouble and expense if you can reasonably "match" the dynamic sensitivity of the device to that which you are measuring. However, the sensitivity matching of the device need not necessarily be 1:1. For example, for a 0.5-1 g signal, a 5 g accelerometer could be used, guaranteeing reasonable sensitivity, little noise contribution by the preamplifiers, and little chance that the transducer would be destroyed if the accelaration levels exceeded 1 g (say 3 g maximum) for even a short time. c.) The frequency response of the measuring device MUST match the overall frequency spectrum to be measured. Every measuring device has a usable frequency band where the device response is optimized (called a "window"). Frequencies to be measured lower than the minimum usable frequency of the device will not be "seen" or measured by the device. On the upper or high frequency side of the

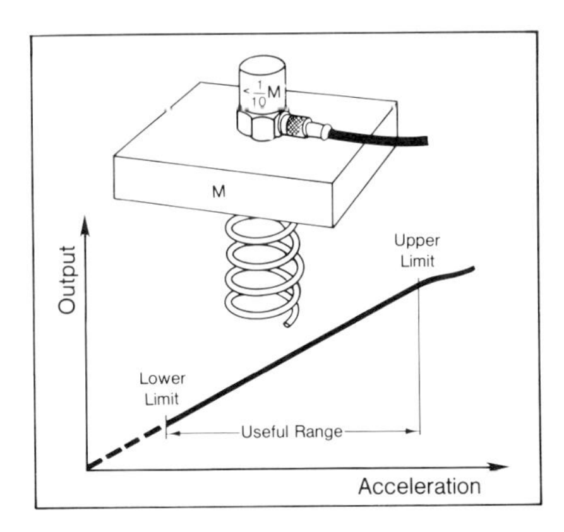

Fig. 4.3 The mass of the measuring accelerometer should be no more than 10% the mass of what is to be measured.
(Courtesy B & K Co.)

window there are two distinct problems: First, if one wants to measure frequencies above the usable window, the device will not see the signal. Second, the accelerometer has its own internal resonance, which when triggered or excited by high frequency vibration within the useable window, will result in an incorrect (high) reading due to the amplification effect created at the device's resonance. It is therefore necessary to electronically insert a low pass filter to guarantee the device resonance does not interfere and exaggerate the high frequency measurement (refer to Fig 4.4). These filters are usually built into the electronics conditioning units. d.) Environmental influences affecting the device are usually characteristic of the device itself. High temperatures drifting of the device outputs are a generic problem to piezoelectric and piezoresistive devices in particular, less so with strain gage devices. e.) Monitoring of the accelerometer in relation to that which is to be measured is critical and will be discussed later in this chapter.

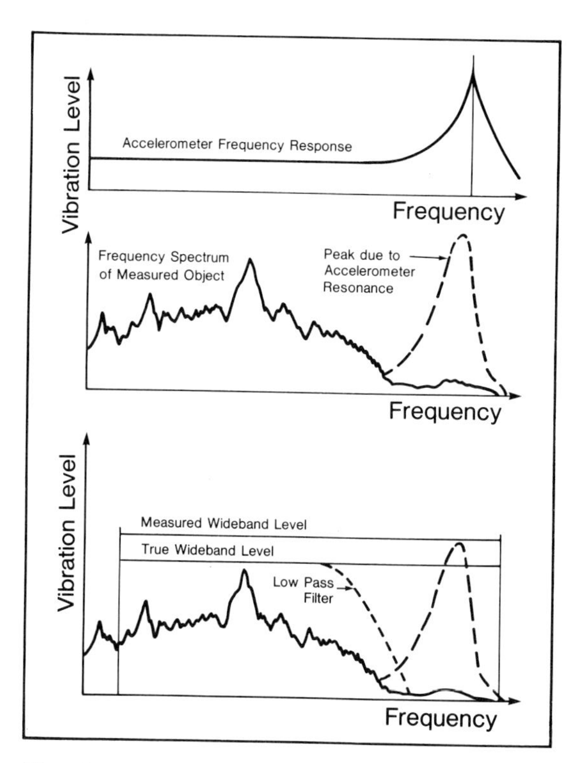

Fig. 4.4 Examples of how a vibrating structure can trigger internal resonances in the measuring accelerometer. Sometimes a low pass filter is used to avoid these resonances.
(Courtesy B & K Co.)

f.) Cross axis sensitivity of an accelerometer is a measure of how well the accelerometer measures vibration in the axis in which it is placed, and rejects vibration in other axis. Thus it is a measure of directionality of the device. Cross axis sensitivity should be zero percent optimally, in reality it is undesirable to use a device whose cross axis sensitivity exceeds 10%. Typical sensitivities are maximally 3-4%.

4.4.1 Piezoelectric crystal accelerometers

Newton's second law states that Force = Mass x Acceleration. The piezoelectric effect states that if one applies mechanical force to crystal material, a voltage results across the crystal face. This voltage is directly proportional to the applied force. The crystal face across which the voltage is developed also develops an electrical charge. This charge is likewise directly proportional to the applied force. If we next take a very small mass and mount it to the crystal and once again apply a varying force, what results is a varying voltage (and charge) across the crystal face which is proportional to the acceleration of the mass moving against the crystal due to the applied force. We have thus created an "accelerometer." Fig. 4.5 shows a cutaway view of two such devices, both of which obey Newton's law. The main differences between the two devices is how the mass and the (crystal) piezo-electric element interact (either in compression or in shear). These devices are available in many sizes, physical configurations, and are designed for multiple uses. The device sensitivity is expressed in terms of coulomb charge per acceleration (either in g units or meters/sec^2). Small mass devices have low charge sensitivity. Larger devices have higher sensitivity. Unfortunately in human measurements, small mass devices are mandatory in order to avoid mass loading inaccuracies and thus there is a sacrifice due to low sensitivity.

Three accelerometers are usually needed to define the three perpendicular linear coordinates which make up most vibration measurements. These three devices are mounted perpendicular (orthogonal) to each other using a single lightweight metal block device (see Fig. 4.6). The entire unit (block plus accelerometers) mass must once again be kept very low to avoid mass loading errors.

Unfortunately one cannot merely connect an electrical recording device directly to the accelerometer (because the crystal signal is small and its output impedance is very high and therefore easily electrically loaded). We therefore need to electronically condition the weak accelerometer signal. First, we must amplify the weak accelerometer signal and simultaneously provide an impedance transformation (high to low) so the signal is adequate for data processing. Part of the conditioning consideration is that if we use a voltage preamplifier we would be extremely limited in cable length between the accelerometer and the preamplifier due to the increase in signal voltage loss as the cable length increases. It turns out that there is another solution to the problem: use an amplifier which senses electrical charge (rather than crystal voltage) and convert the signal into electrical voltage after it has been amplified. This is known as a "charge amplifier" which is described next.

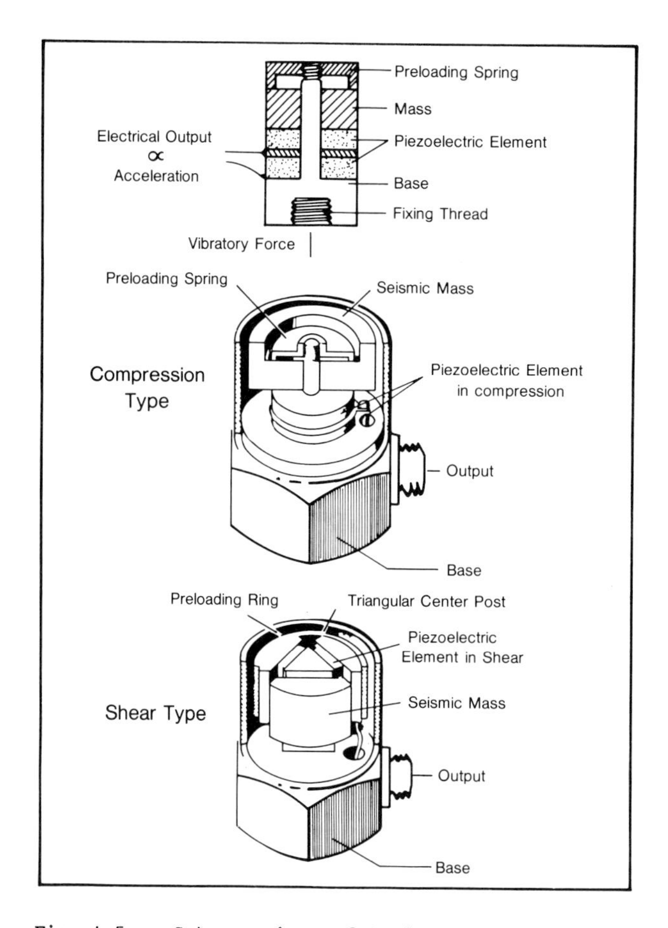

Fig. 4.5 Cutaway views of typical B & K compression and shear type accelerometers.
(Courtesy B & K Co.)

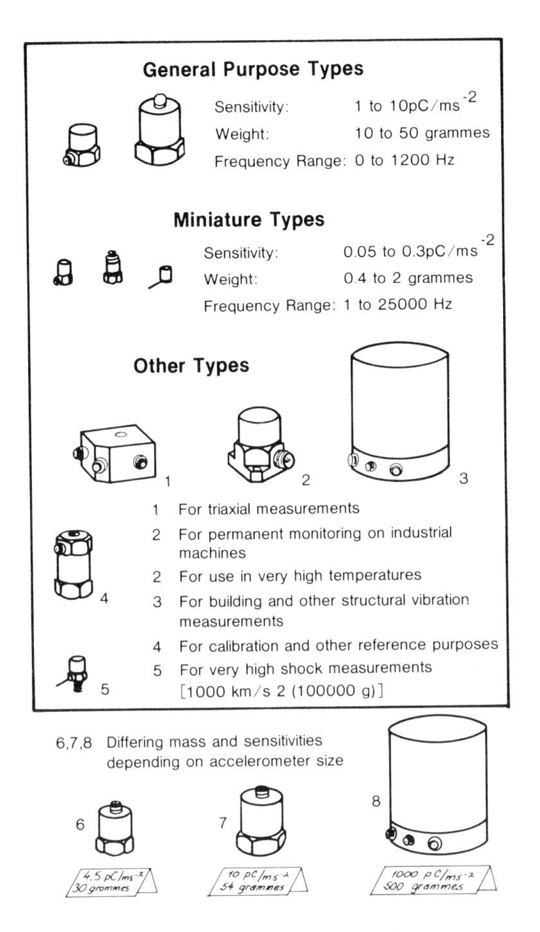

Fig. 4.6 Various piezoelectric accelerometers and their typical uses. (Courtesy B & K Co.)

4.4.1.1 Charge Amplifiers (7,8)

A simplified equivalent electrical circuit of a piezoelectric accelerometer is shown in Fig. 4.7.

The charge Q is equal to the open circuit voltage E multiplied by the accelerometer capacity C_a:

$$Q = C_a \times E \tag{1}$$

As previously stated, the output charge generated is proportional to the acceleration impressed upon the accelerometer. Furthermore, the charge generated remains constant regardless of the amount of external capacitance added to the transducer by long signal wire lengths.

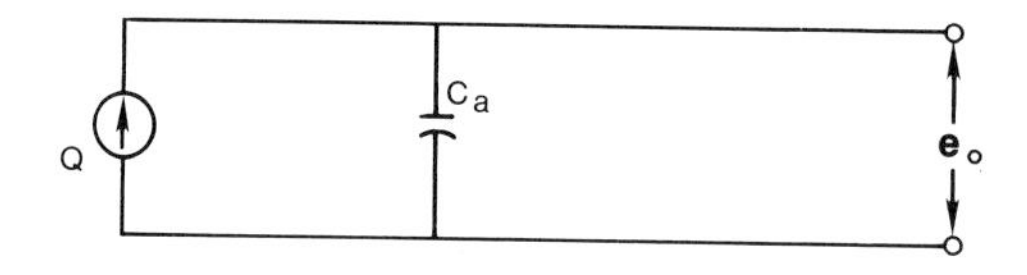

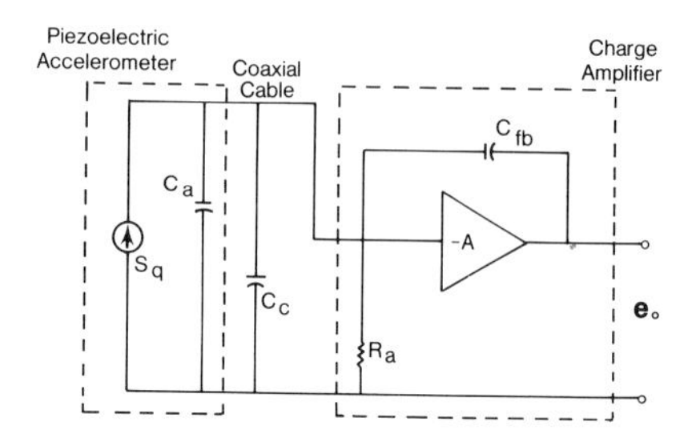

Fig. 4.7 Piezoelectric accelerometer equivalent electrical circuit. (Courtesy Columbia Research Co.)

Fig. 4.8 Representative equivalent circuit of piezoelectric accelerometer, connecting cable, and charge amplifier. (Courtesy Columbia Research Co.)

The functional circuit of an accelerometer charge amplifier system is shown in Fig. 4.8 In this amplifier the capacitive component of the input impedance is very large. The amplifier derives its high capacitive component by negative feedback and is proportional to the open loop gain "A" of the amplifier. Hence, the effective input capacitance of the amplifier is:

$$C = C_{fb}\,(1 + A) \tag{2}$$

The overall charge gain A_q of the amplifier is defined as,

$$A_q = \frac{-E_o}{E_t C_t} \tag{3}$$

and the charge gain of the amplifier in terms of volts output/picocoulomb input is,

$$A_q = \frac{1}{C_{fb}} \; \frac{1}{1 + \frac{1}{A} + \frac{C_c + C_t}{A\,C_{fb}}} \tag{4}$$

normally,

$AC_{fb} > C_c + C_t$ and $A \gg 1$.

Therefore, equation (4) becomes approximately,

$$A_q = \frac{1}{C_{fb}} \tag{5}$$

But the amplifier's output is obtained as a voltage which is a virtual measure of the amplified voltage across the feedback computer C_{fb}. Thus, the output of the charge amplifier has transformed a measurement of charge (which is proportional to acceleration of the mass against the crystal) to a corresponding voltage, at low impedance (e.g., output impedance of operational amplifiers on normally low). All of this means, that the coaxial cable length from the accelerometer is independent of and does not alter the charge amplifier's output, only the actual acceleration signal varies the charge amplifier output, which is what is desired.

The accuracy with which a piezoelectric accelerometer responds to a vibratory motion is of considerable importance. The upper frequency response is usually limited by the first mechanical resonance of the accelerometer system and is to be determined from calculations based on mechanical characteristics. In this configuration, usually the lower frequency response is limited by the electrical parameters of the conditioning amplifier and the instrumentation that follows. Consequently, the lower frequency response is generally determined from the electrical considerations. In particular, the low frequency response of a crystal accelerometer is related to the resistance multiplied by the capacitance, forming the (RC) time constant of the accelerometer and the conditioning electronics. Briefly, the lower cut-off frequency f (-3 db point) is given by:

$$f = \frac{1}{2\pi RC} \qquad (6)$$

where, C = the sum of the crystal capacitance, cable capacitance, and amplifier input capacitance ($C_c + C_s + C_a$) in farads.

R = the input resistance of the conditioning amplifier in ohms.

Equation (6) shows that maximum low-frequency response is obtained with a large input resistance-capacitance RC product. This high input impedance is easily obtained with integrated circuit operational amplifiers which use Field Effect Transistor (FET) input devices.

The charge amplifier's output normally goes to one channel of an FM tape recorder. Fig 4.9 shows a typical triaxial measurement arrangement using three crystal accelerometers. A four channel FM tape recorder is used for high quality recording of the data, one of the four channels is used either as a voice track or for a high frequency sinusoid reference (100,000 Hz) to minimize tape recorder motor wow and flutter fluctuations. One uses an FM tape system for fidelity and to guarantee that the Fourier spectrum obtained from the data is accurate. Poor tape recorder wow and flutter can cause erroneous spectra when analyzing vibration data. Namely, spectral peaks appear which were not part of the original data. Thus the quality of the tape recorder used is critical to a final analysis of the data. The monitoring oscilloscope is used to guarantee that the charge amplifier(s) are not being overloaded and saturating. If the tape recorder should record these preamplifier overload signals then once again the spectrum analysis will be inaccurate and misleading in the low frequency part of the spectrum (known as "DC shifting").

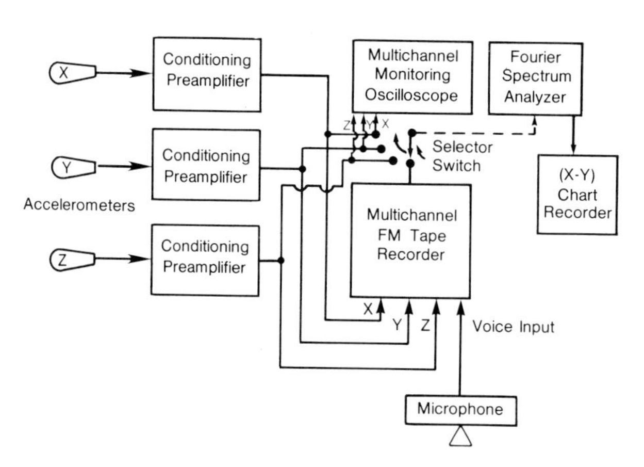

Fig. 4.9 Typical triaxial acceleration measurement system block diagram.

The Influence of the Environment on Crystal Accelerometer

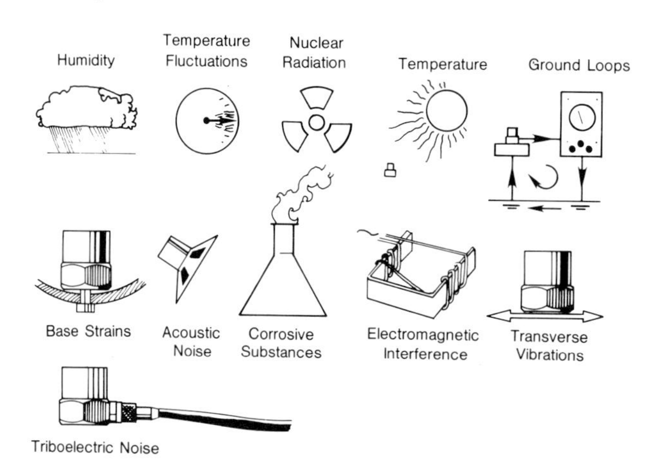

Fig. 4.10 Sources of environmental influences on piezoelectric accelerometers.
(Courtesy B & K Co.)

Piezoelectric (crystal) accelerometers are used principally for hand-arm vibration tool measurements, not whole-body measurements (because they do not have DC response). In tool measurements high acceleration levels spanning a wide frequency band (6 Hz - 2000 Hz) are expected. These devices are generally reasonably rugged and to some extent protected from overload acceleration levels.

Piezoelectric crystal accelerometers are influenced by temperature, humidity, etc. as shown in Fig. 4.10 and are briefly described next (9).

Temperature - Typical general purpose crystal accelerometers can tolerate temperatures up to 250°C. At higher temperatures the piezoelectric material looses its sensitivity will be permanently altered. Such an accelerometer may still be used after recalibration if the damage is not too severe. For temperatures up to 400°C, accelerometers with a special piezoelectric ceramic are available.

All piezoelectric materials are temperature dependent so that any change in the ambient temperature will result in a change in the sensitivity of the accelerometer. For this reason most manufacturers of accelerometers provide a sensitivity versus temperature calibration curve so that measured levels can be corrected for the change in accelerometer sensitivity when measuring at temperatures significantly higher or lower than 20°C.

Piezoelectric accelerometers also exhibit a varying output when subjected to small temperature fluctuations, called temperature transients, in the measuring environment. This is normally only a problem where very low level or low frequency vibrations are being measured. Modern shear type accelerometers have a very low sensitivity to temperature transients.

When accelerometers are to be fixed to surfaces with higher temperatures than 250°C, a heat sink and mica washer can be inserted between the base and the measuring surface. With surface temperatures of 350 to 400°C, the accelerometer base can be held below 250°C by this method. A stream of cooling air can provide additional cooling.

Electrical - Since piezoelectric accelerometers have a high output impedance, problems can sometimes arise with electrical noise signals are induced in the connecting cable. These disturbances can result from ground loops, triboelectric noise or electromagnetic noise (see Fig. 4.11). Ground loop currents sometimes flow in the shield of accelerometer cables because the accelerometer and measuring equipment are grounded at different points in the circuit. The ground loop is broken by electrically isolating the accelerometer base from the mounting surface by means of an isolating stud and mica washer as already mentioned. Tribo-electric noise is often induced into the accelerometer cable by mechanical motion of the cable itself. It originates from local capacity and charge changes due to dynamic bending, compression and tension of the layers making up the cable. This problem is avoided by using a proper graphited accelerometer cable and taping or gluing it down as close to the accelerometer as possible. Electromagnetic noise is often induced in the accelerometer cable when it lies in the vicinity of running machinery operated from power lines. Double shielded cable helps in this respect, but in severe cases a balanced accelerometer and differential preamplifier should be used to avoid power line interference.

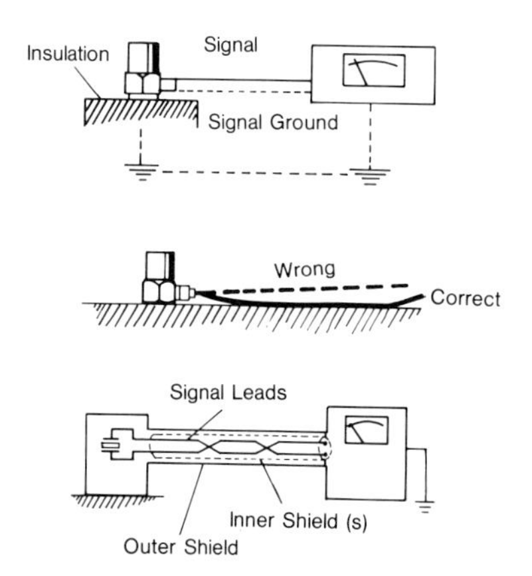

Fig. 4.11 Sources of electrical interference and ground loops on measurements using piezoelectric accelerometers. (Courtesy B & K Co.)

Other influences - (see Figs. 4.12 and 4.13) - Base strains: When an accelerometer is mounted on a surface that is undergoing strain variations, an output will be generated as a result of the strain being transmitted to the sensing element. Most accelerometers are designed with thick, stiff bases to minimize this effect. Nuclear Radiation: Many accelerometers for example B&K devices can be used under moderate gamma radiation doses without significant change in their characteristics. Contact the manufacturer of the accelerometer which you are using for specifics. Magnetic Fields: The magnetic sensitivity of piezoelectric accelerometers is very low, normally less than 0.01 to 0.25 m/s^2 per k Gauss in the least favorable orientation of the accelerometer in the magnetic field. Humidity: Most accelerometers are sealed to ensure reliable operation in humid environments. For short duration use in liquids, or where heavy condensation is likely, Teflon sealed accelerometer cables are recommended. The accelerometer connector should also be sealed with an acid free room temperature vulcanizing silicon rubber or

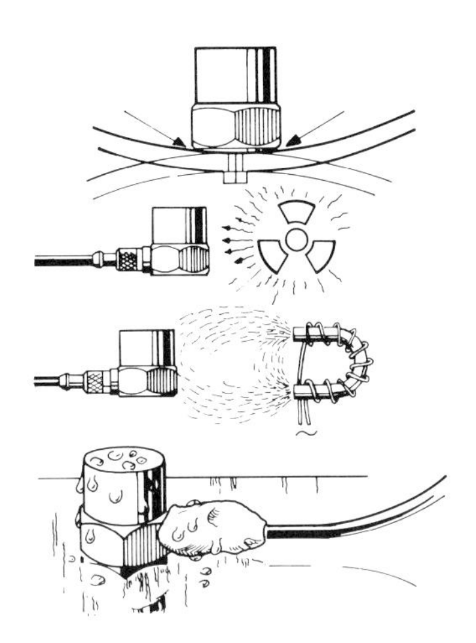

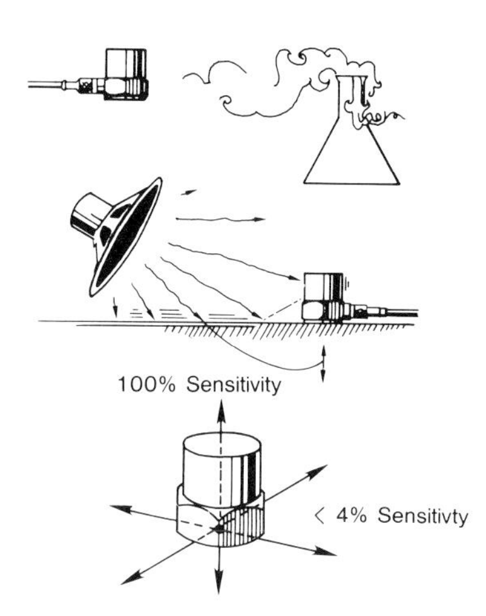

Fig. 4.12 Additional sources of environmental influences on piezoelectric accelerometers.
(Courtesy B & K Co.)

Fig. 4.13 Chemical, acoustic, and cross axis sensitivity influences on piezoelectric accelerometers.
(Courtesy B & K Co.)

mastic. Industrial accelerometers with integral cables should be used for permanent use in humid or wet areas. Corrosive Substances: The materials used in the construction of accelerometers usually have a high resistance to most of the corrosive agents encountered in industry. Acoustic Noise: The noise levels present in machinery are normally not sufficiently high to cause any significant error in vibration measurements. Normally, the acoustically induced vibration in the structure on which the accelerometer is mounted is far greater than the airborne excitation. Transverse Vibrations: Piezoelectric accelerometers are sensitive to vibrations acting in directions other than coinciding with their main axis. In the transverse plane, perpendicular to the main axis, the cross sensitivity is less than 3 to 4% of the main axis sensitivity (typically < 1%). As the transverse resonant frequency normally lies at about 1/3 of the main axis resonant frequency, it should be considered where high levels of transverse vibration are present.

4.4.1.2 Piezoelectric Accelerometer Calibration

One of the shortcomings of crystal accelerometers is that they do not have DC (or zero Hz) response and therefore are not self calibrating. One must vibrate the accelerometer with a sinusoid and compare its output at various frequencies to a calibrated accelerometer. Initially accelerometers are individually calibrated at the factory and the calibration sheets are shipped with each unit (Fig. 4.14). They must be consulted before using the device. Where accelerometers are stored and operated within their specified environmental limits, i.e. are not subjected to excessive shocks, temperatures, radiation doses, etc. generally, there will be a minimal change in characteristics over a long time period.

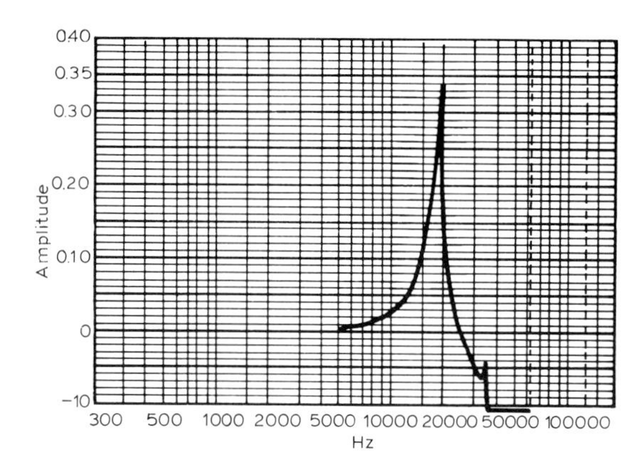

Fig. 4.14 Typical piezoelectric calibration chart provided by the accelerometer manufacturer. (Courtesy B & K Co.)

However, in normal use in the workplace, accelerometers are often subjected to unavoidably harsh treatment which can result in a significant change in characteristics and sometimes even permanent damage. When dropped onto a concrete floor from hand height an accelerometer can be subjected to a shock of many thousands of g's. It is wise, therefore, to treat these devices very carefully and periodically check their sensitivity calibration. This is normally sufficient to confirm that the accelerometer is not damaged. B&K for one has introduced a battery-powered calibrated vibration source as a convenient means of performing accelerometer calibrations (Fig. 4.15). This calibrator has a small built-in shaker table which can be adjusted to vibrate sinusoidally at precisely 1 g = 9.81 m/s^2.

The sensitivity calibration of an accelerometer is checked by fastening the device to the shaker table and noting its output when vibrated at 1 g. Alternatively one known accelerometer can be used as a reference device for another device. It is mounted on the shaker table with the accelerometer to be calibrated. One can note their respective outputs on a multi-channel oscilloscope. Since the sensitivity of the reference accelerometer is known, the unknown accelerometer's sensitivity can then be determined.

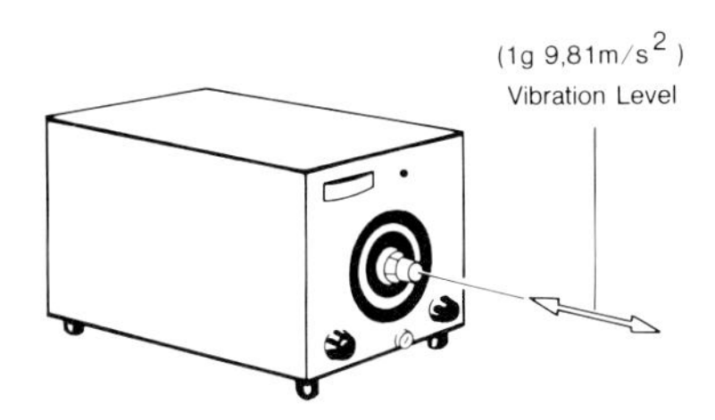

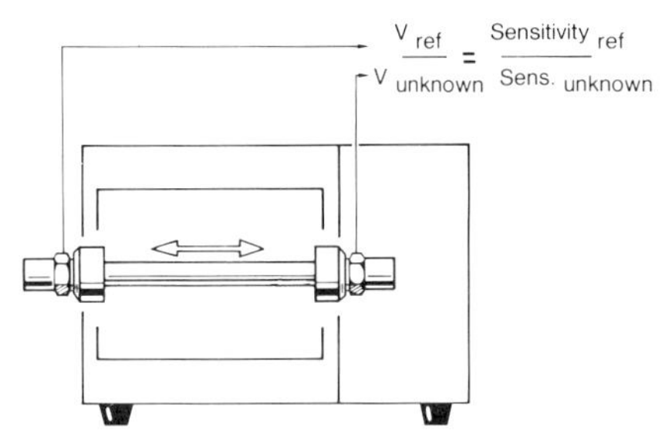

Fig. 4.15 Typical piezoelectric portable calibrator containing a known internal reference accelerometer which is compared to the accelerometer under test.
(Courtesy B & K Co.)

An equally useful application of a portable calibrator is the checking of a complete measuring or analyzing setup before the measurements are made. The measuring accelerometer is simply transferred from the measuring object to the calibrator and vibrated at a level of 1 g. The entire measurement system including the final readout can be checked. If a tape recorder is being used, the 1 g calibration level can be recorded before measurements have begun.

4.4.2 Piezoresistive Accelerometers

Piezoresistive accelerometers are the accelerometer of choice for human whole-body vibration measurements. Their operations and characteristics are very different than piezoelectric devices. Fig. 4.16a shows the operating principles. Four pieces of either P or N type semiconductor form the four arms of a Wheatstone bridge. A small metal beam is sandwiched between each pair of semi-conductor materials. At the other end of the beam is affixed a small mass "m." Once again by Newton's law, we measure the acceleration of the mass due to force f applied to it through the bending motion of the metal beam. If we initially balance the Wheatstone bridge with no force on it, when vibration is applied to the accelerometer the moving beam unbalances he bridge by a precise amount. The millivolt electrical output from the unbalanced bridge is next directed to a high gain differential (D.C.) operational amplifier which amplifies the small signal and conditions it for use by a tape recorder, Fourier analyzer, etc. Piezoresistive devices have several advantages and a few disadvantages: a.) The device is self calibrating because it has zero Hz or D.C. response, b.) Its useable frequency response extends from 0 Hz - 500 Hz, and sensitivities are about 12 millivolt/g output, c.) It is very light weight (1-2 grams), thus minimizing mass loading problems for human work, d.) It requires DC-bridge excitation and a stable DC differential amplifier, e.) The device must be temperature compensated and must be mechanically damped (or else it will electrically ring or oscillate).

Overall the device is excellent for human whole-body vibration work. These requirements usually are: a.) Frequency response 0 Hz - 80 Hz, b.) Acceleration levels rarely exceed 1-2 grms, c.) Direct calibrated, d.) Very light weight, e.) Accelerometer should be about 0.7 critical damping.

The device can be calibrated in two distinct ways: a.) placing it on a portable calibrator (or similar shaker device), and/or b.) self calibrate the unit.

4.4.2.1 Self Calibration of Piezoresistive Accelerometers

The calibration procedure is generally as follows: Attach the device to a suitable DC voltage (usually 12-15 VDC) for bridge excitation and hook up differential amplifier. Balance the amplifier output for zero output with the "sensitive surface" vertical (facing the force of gravity). Rotate the device 90 degrees thus removing the force of gravity from the sensitive surface. The voltage deflection obtained is exactly equal to 1 g and is incrementally linear. Record the magnitude of the deflection. This deflection is the desired calibration. Return the device vertically to its original operating position for zero output. Measurements can now be made and compared in amplitude to the calibration. It is very desirable to have a calibration at

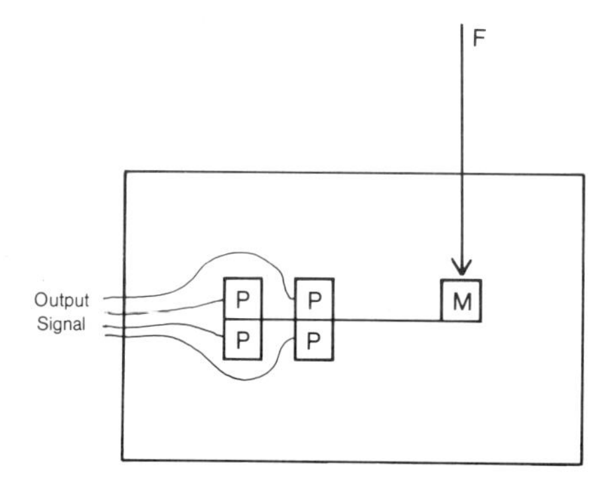

A. Internal construction

Fig. 4.16 (a) Operational diagram of a typical lightweight miniature piezoresistive accelerometer, (b) view of an actual ENTRAN device compared to a one cent coin.
(Courtesy ENTRAN Co.)

the beginning, middle, and end of each measurement since the device may tend to drift occasionally.

4.4.3 Strain Gage Accelerometer

Strain gage accelerometers are usually used in conjunction with whole-body measurements, on vehicle floors, tops of shaker tables, etc. when the mass of the structure to be measured are large, because strain gage accelerometers are usually heavy and bulky. Operationally, four wires in the device form the arms of a Wheatstone bridge. If these wires are compressed or stretched, their electrical resistance changes in proportion to the compression or stretching, thus unbalancing the Wheatstone bridge. Once again, if a small mass is placed against these wires and vibration applied, we obtain the acceleration of this mass due to the vibration force. Signal conditioning for this device is virtually the same as for the piezoresistive accelerometers. Self calibration is also identical. However, one cannot easily use the previously mentioned portable calibrator with strain gage accelerometers because of their large mass. As a rule these devices are usually thermally stable and are low drift devices. In many whole-body studies it is possible to use an appropriate mix of strain gage and piezoresistive devices, the former on large structures, the latter on the worker. Strain gage device frequency response is typically 0-50 Hz. Some units will go as high as 100 Hz in response. These devices produce reasonably high voltage/g sensitivity because of their large mass. The mounting of these devices is similar to piezoresistive accelerometers. A typical device is shown in Fig. 4.17.

4.5 MECHANICAL IMPEDANCE MEASUREMENTS

Mechanical impedance is defined as the ratio of force to resulting velocity of a system being vibrated. Both measurements can be taken at the same point (called point impedance) for light structures or for large structure forces and velocity are obtained at different points (called transfer impedance). Occasionally in human vibration work it is necessary to make point impedance measurements, especially when trying to obtain resonances (10,11). The point impedance transducer is really two devices in one package: a.) an accelerometer, and b.) a force cell. Since we seek both force and velocity measurements, an accelerometer is used to measure acceleration, and then by a single electronic integration the velocity function is obtained. The force cell is merely the transducer without a mass, thus force can be obtained by letting vibration impinge directly on an accelerometer <u>without</u> an internal mass. (The reader will recall that acceleration was obtained by letting vibration impinge on a small internal mass which in turn was in contact with either a crystal

Fig. 4.17 Typical strain gage accelerometer.
(Courtesy Stratham Co.)

element, or on a small beam, or pressed against wires whose resistance changed.) A typical crystal "impedance head" is shown in Fig. 4.18. In Fig. 4.18a the vibration force impinges directly on a crystal element producing an electrical output. In Fig. 4.18.b both the crystal accelerometers and the force cell are "sandwiched" together as one, with both force and acceleration outputs as shown. The instrumentation used with force signals is the same as that used with other accelerometers.

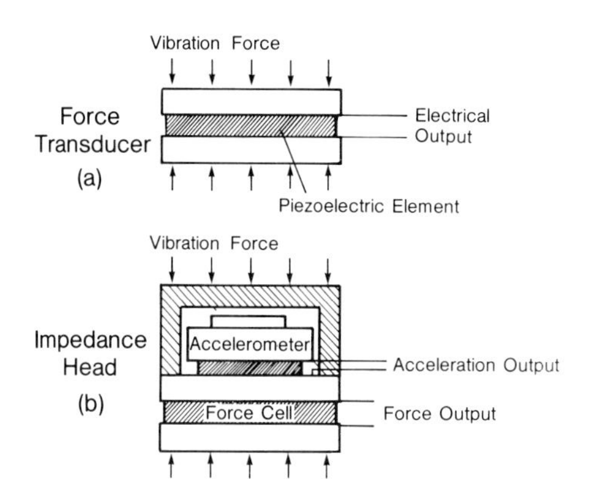

Fig. 4.18 Typical mechanical impedance transducer consisting of a force cell (a) combined with a piezoelectric accelerometer (b).

4.6 ACCELEROMETER MOUNTING FOR VIBRATION STUDIES

Proper mechanical mounting of the accelerometer is absolutely critical for a accurate measurement. In human vibration work mounting the device takes two forms: a) Mounting to the vibration source (e.g. hand-tool, vehicle seat, etc.) and b) Mounting to the human body.

4.6.1 Mechanical Mounting of the Accelerometer to the Vibration Source (8,9)

Proper mounting of an accelerometer means the transducer is rigidly affixed to the vibrating source and in line with the direction of vibration. If the accelerometer is not rigidly affixed it can rattle producing a false output signal resulting in false spectra. With poor mounting, there is a tendency to lower the natural frequency of the system thereby reducing the effective band width (window) of the device. If the device is not in line with the vibration motion a part of the measurement will be lost due to the misalignment. Many accelerometers come with threaded studs which can be screwed directly into for example a vehicle floor. An alternative mounting method is to use a thin layer of beeswax between the accelerometer and the test structure (limited to

40°C in temperature). An epoxy or cyanoacrylate cement can also be used, but not a soft glue (because soft material acts as a mechanical low pass filter). When electrical isolation between the accelerometer and the test structure are required, a thin mica washer or isolated stud can be used. In some instances the accelerometer can be mounted to a strong permanent magnet which in turn is affixed to a metal test structure. (This technique however is not to be used as a first choice.) Holding the accelerometer against the test structure is the least desirable way to make a measurement since erroneous and nonrepeatable results can be expected. An excellent method of mounting an accelerometer is to use stiff Mylar-type double sided carpet tape. Wipe the surface to be measured free of dirt, grime and moisture; place a piece of carpet tape on the structure with one sticky side. Expose the other surface and firmly affix the accelerometer to it. Take a second strip of tape and place each end on the vibrating surface by tautly holding the top of the accelerometer against the surface to be measured (see Fig. 4.19).

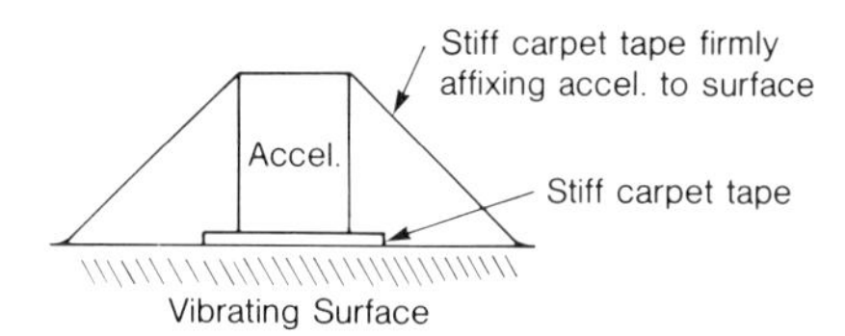

Fig. 4.19 Accelerometer mounting technique using stiff double sided carpet tape.

In the case of vibrating pneumatic hand-tools where acceleration is expected to be very high other mounting techniques need to be used and have been developed. Fig. 4.20 shows an automotive hose clamp, with a welded steel cube affixed. This cube is tapped and threaded in three perpendicular directions. One or more piezoelectric accelerometers are mounted to the cube. The hose clamp is then firmly affixed to the tool handle where the workers grasp the tool (5,10) and measurements are obtained. Fig. 4.21 shows a method for mounting an accelerometer to a chipping hammer chisel (where the worker holds and guides the tool). Since these acceleration levels are very high, it was necessary to actually weld the mounting block directly to the chisel (10) and then provide a special retainer to prevent the accelerometer from detaching itself from the mount. (The reader will note only one accelerometer is used here, the reason is that the potential for accelerometer destruction is very high, and since accelerometers are quite expensive we did not wish to chance destruction of three units at once. We therefore chose to obtain measurements

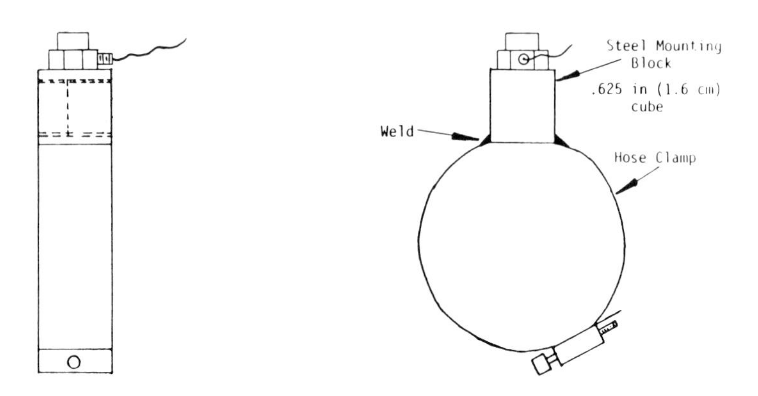

Fig. 4.20 Hose clamp/accelerometer mounting block used for some hand-tool measurements.
(Adapted from Wasserman et al, reference (10))

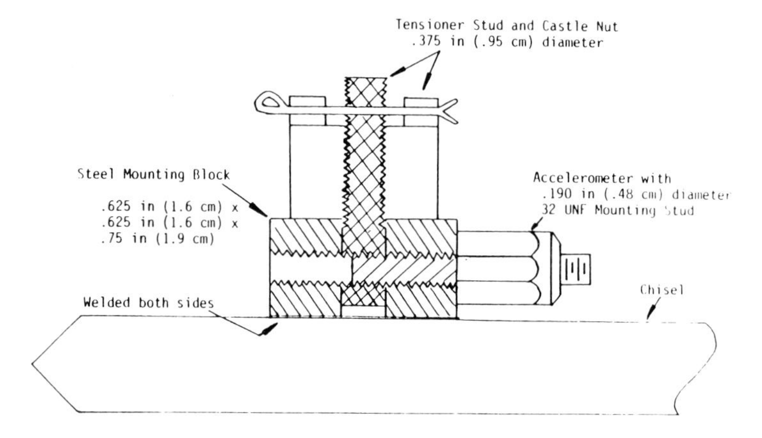

Fig. 4.21 Accelerometer mounting block used on chipping hammer chisels.
(Adapted from Wasserman et al, reference (10))

along the principle axis of vibration, realizing that although the other two axis contributed to the vibration, in this case their contribution was small in comparison to the principle axis). Fig. 4.22 shows the actual mounting of the various fixtures of Fig. 4.20 and 4.21 to a pneumatic chipping tool. Figs. 4.23 and 4.24 schematically depicts the hose-clamp accelerometer mounting arrangement for two types of pneumatic grinders.

Two of the difficult problems with making measurements from vibrating tools are a.) potential destruction of the accelerometer and b.) DC or zero shifting (3-5); the former can be solved by either using very rugged and expensive accelerometers (e.g. B&K 8309) or using a mechanical filter whose

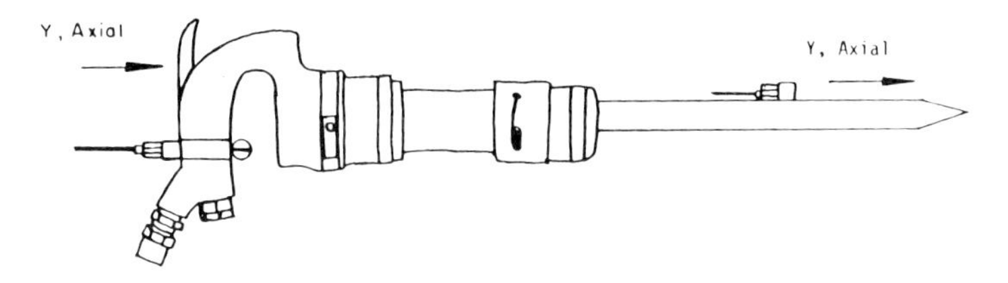

Fig. 4.22 Pneumatic chipping hammer acceleration measurement and coordinate system on both the chisel and rear handle.
(Adapted from Wasserman et al, reference (10))

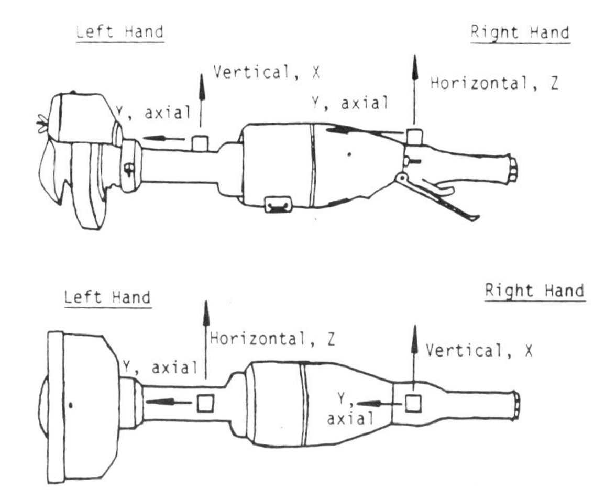

Fig. 4.23 Pneumatic horizontal grinder acceleration measurements and coordinate system.
(Adapted from Wasserman et al, reference (10))

filter characteristics does not alter the acceleration range of interest (approx. 6-2000 Hz); the latter produce a false low frequency spectrum due to preamplifier overload, and saturation, thus each vibration must be constantly monitored using an oscilloscope at each of the preamplifier outputs.

In the case of whole-body vibration (when using piezoresistive accelerometers) and measuring vibration transmitted through a vehicle seat impinging on the buttocks of a driver, a triaxial accelerometer/metal cube arrangement can also be used. The accelerometer is imbedded to the center of a hard-rubber disk as shown in Fig. 4.25(a). Access to the accelerometers and their corresponding wires is via a thin metal mounting disc as shown. In practice, each accelerometer is first calibrated using the rollover method. Next, the instrumentated disc is placed on the seat with the accelerometers positioned where the operator's buttocks would be located and with the metal mounting disc facing downward into the seat cushion. Next, place tape only on

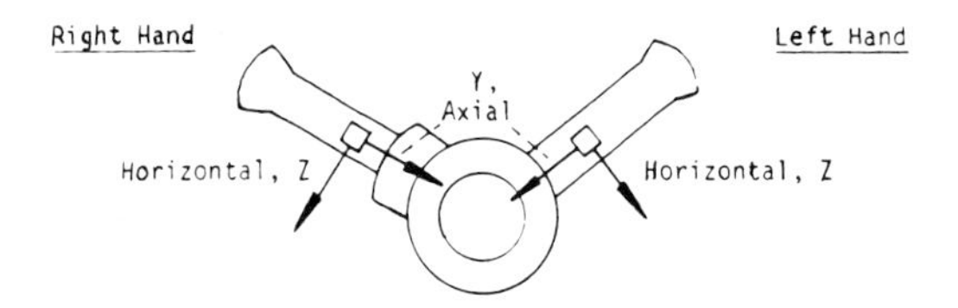

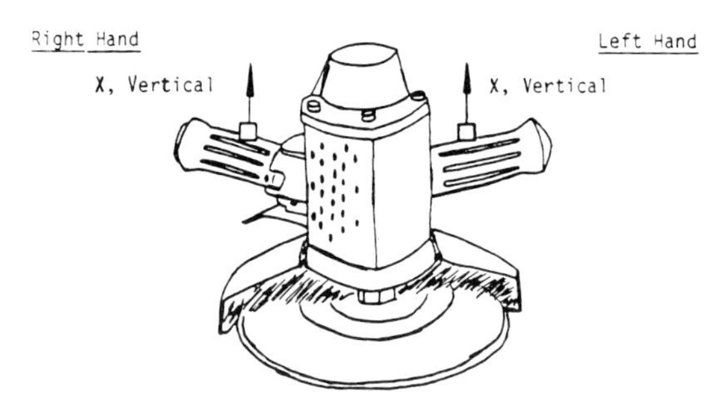

Fig. 4.24 Pneumatic vertical grinder acceleration measurements and coordinate system.
(Adapted from Wasserman et al, reference (10))

the perimeter of the disc to the seat cushion, avoiding any tape which might cover the center of the disc where the accelerometers are located. Finally, strain relieve the wires coming from the disc center by taping to the seat cushion or other parts of the seat structure (Fig. 4.25(b)).

When the floor of the vehicle is to be instrumentated either large piezoresistive or strain gage accelerometers can be used and affixed directly to the floor (either using double sided carpet tape, gluing, etc. as appropriate) at the points where the operator's feet are located. Similarly, as with hand-arm measurements, a hose-clamp/accelerometer arrangement can be used for mounting to steering wheels, gear levers, etc.

4.6.2 Attaching Accelerometers to the Human Body

It is virtually impossible to easily and noninvasively mount even light weight accelerometers to the human body. Although it cannot and should not be done morally and ethically, optimally, it would be desirable to directly mount accelerometers to bony areas prominances of the body (e.g. metacarpals, wrists, elbows for hand-arm measurements, the skull, jaw, spinal column, knee, shoulder, etc. for whole-body measurements). Mounting accelerometers to areas of soft tissue (e.g. stomach) is not desirable for two reasons: a.) It is difficult to clearly anatomically locate a given point and describe where the accelerometer(s) were mounted; b.) where soft tissue is primarily present,

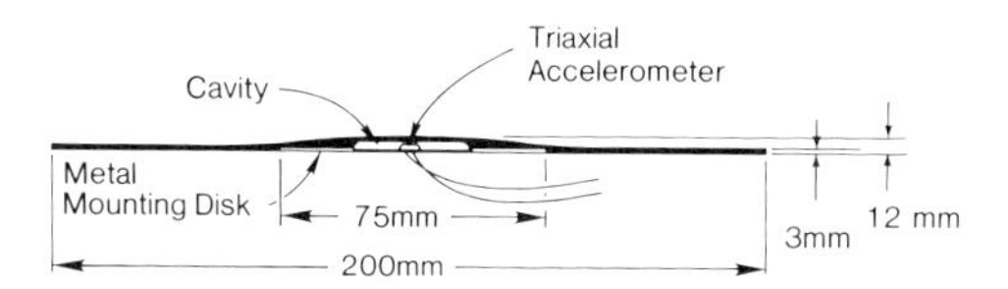

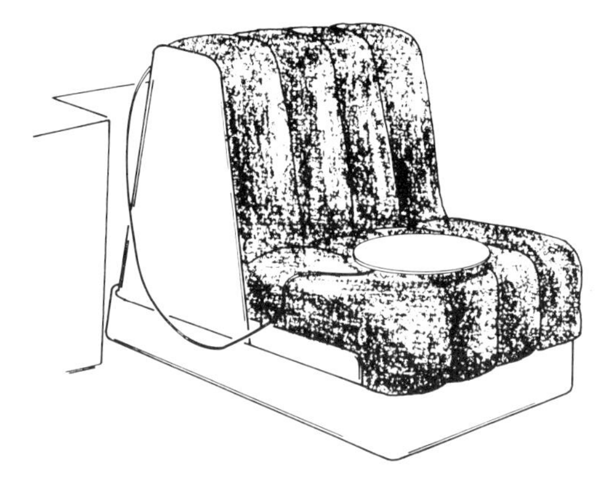

Fig. 4.25 (a) seat disk accelerometer arrangement used to measure vibration transmitted to a vehicle operator; (b) unit in actual use. (Adapted from Wasserman et al, references (13,14))

there are multiple tissue layers "sliding" due to the vibration impinging on the body. The net result of all this is one really never knows just what one is measuring when working with soft tissue.

Unfortunately, it is not possible nor ethical to mount accelerometers directly to bone, in a living system. Thus, one is limited to mounting accelerometers noninvasively (i.e. not piercing the skin) near bony surfaces and prominances. Then the question becomes, what data is lost from a surface mounted accelerometer measurement with respect to an accelerometer actually mounted to the bone? The answer is insignificant loss providing the accelerometer weighs one or two grams or less. This has been shown in a classic study by Ziegart and Lewis (12) who compared vibration data obtained by a light weight 1.5 g accelerometer/strap device mounted around the lower leg (near the calf) of volunteer human subjects. Another 1.5 g lightweight accelerometer was rigidly mounted to the lock end of a hypodermic needle. A small area near the first accelerometer was anesthetized and the needle/accelerometer arrangement was inserted against the femur bone. The leg was vibrated and both accelerometer outputs were recorded and compared. They were virtually identical. However, in a repeat experiment, when two 34 g accelerometers whose mass' were larger than the 1.5 g device, great differences resulted between the 34 g devices. These were attributed to mass loading and tissue resonances (see Figs. 4.26 and 4.27).

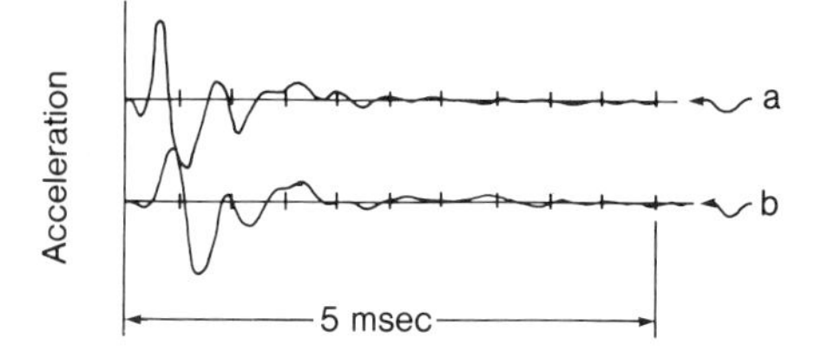

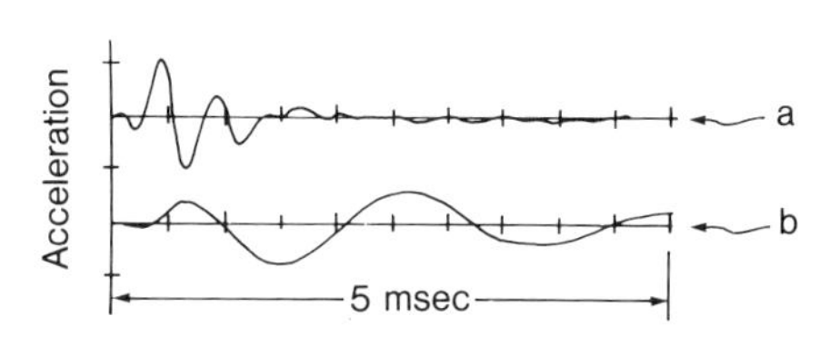

Fig. 4.26 Comparison of actual bone acceleration to skin acceleration using 1.5 g accelerometers, (a) needle mounted accelerometer used with bone, (b) skin mounted accelerometer.
(Adapted from Ziegart et al, reference (12))

Fig. 4.27 Comparison of actual bone acceleration to skin acceleration using 34 g accelerometers, (a) needle mounted accelerometer used with bone, (b) skin mounted accelerometer.
(Adapted from Ziegart et al, reference (12))

When attempting to measure whole-body vibration researchers have been very creative. When measuring vibration at the cranial level, a taut band around the skull containing small accelerometer(s) can be used, or a metal bite bar (developed by the U.S. Air Force AMRL and shown in Fig. 4.28) can be used with from one to three miniature accelerometers to measure vibration at the workers jaw (3,14). In this case a simple dental impression is made by using warm dental material on one end of the bite bar. The subject bites once into the warm material, it hardens, and the device can then be used for measurements.

Usually, when accelerometers are affixed at bony protusions (i.e., knee, shoulder blade, etc.) then stiff double-sided Mylar type carpet tape is used (see Fig. 4.19). If one uses a soft or spongy tape, then the tape itself can act as a mechanical filter between the skin surface and the accelerometer and possibly filter out the high vibration frequencies impinging on the worker. A stiff tape will be mechanically transparent to the higher frequencies and will allow the accelerometer to actually measure all of the impinging vibration (13,14).

When attempting to measure hand-arm vibration it is virtually impossible to easily mount accelerometers to the metacarpals. Rasmussen has devised a device (15) whereby a small metal unit is held against the vibrating hand tool but yet retained between the metacarpals by the worker in an attempt to actually avoid placing or taping an accelerometer directly to the metacarpal (sweating, grime, etc. will not allow for long measurements). Pyykko (16) has used a taut wrist-band type device to measure vibration at the wrist. Mostly, however, researchers mount accelerometers to the vibrating tool handles were the workers grip the tools. Sometimes attempts are made to determine the workers' grip strength on the handles in an effort to calculate the mechanical coupling factor to the hands (i.e. how much of the total vibration reaches the hands as a function of hand coupling). Since the lighter the grip the less vibration getting into the hands and fingers, the more the grip, the more energy reaching the hand and fingers (3-6).

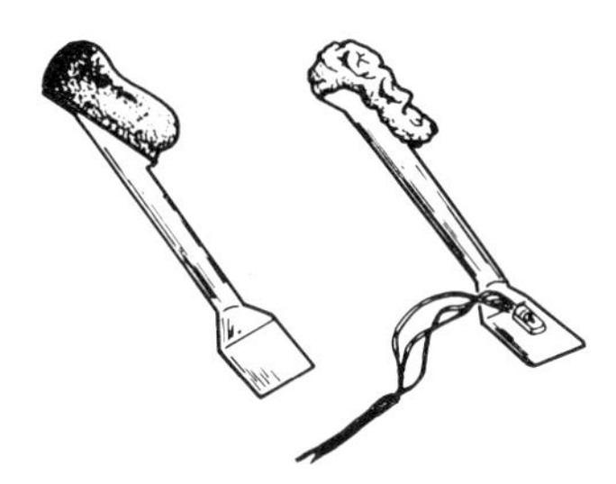

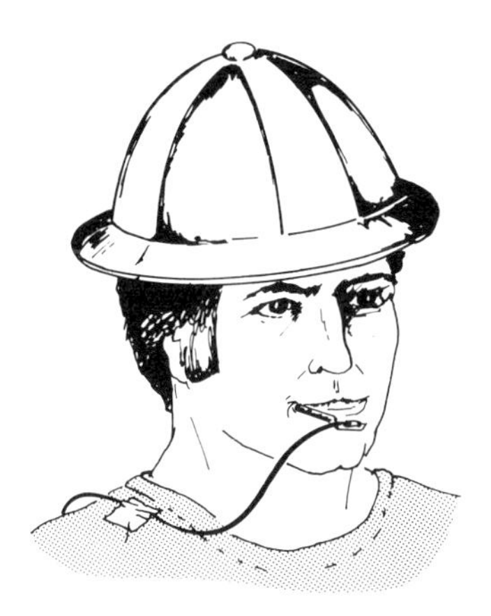

Fig. 4.28 Brass bitebar device used to measure cranial level acceleration. (upper left) bitebar alone, (upper right) bitebar with dental impression and accelerometer, (lower) device being used for workplace and measurements.
(Adapted from Wasserman et al, references (13,14))

4.7 VIBRATION INSTRUMENTATION CONFIGURATIONS

The reader will recall that the basic instrumentation configuration for obtaining occupational vibration measurements were the following common elements: a.) an accelerometer, b.) a corresponding conditioning preamplifier, c.) a multichannel FM tape recorder, d.) A multichannel monitoring oscilloscope, and (optionally), e.) a Fourier Analyzer with its corresponding (X-Y) chart recorder for spectral plots. Needless to say, as one increases the number of vibration channels, the equipment grows proportionately (e.g. more preamplifiers, increased channel capacity in the FM tape recorder and monitoring oscilloscope). Similarly, decreasing the number of vibration channels reduces the equipment needs.

There are additional equipment configurations which can be considered too. Referring to Fig. 4.29 and using trial measurements as one arrangement we can add three new elements to Fig. 4.19: a weighting filter, rms unit, analog or digital meter readout. This configuration can or can not be used with the Fourier Analyzer and chart readout. What we seek with the addition of these three elements is a single number measurement which can be compared to a human vibration standard. When we use an electronic weighting filter its band-pass window is shaped as the <u>reverse</u> or mirror image of the standard curve which we compare it against. For example, Fig. 4.30 is a weighting filter used in hand-arm vibration standard ISO5349 and the ACGIH hand-arm standard (16) (to be discussed in the next chapter). The solid-continuous line represents the inverse (or bandpass) of the filter. Filter gain is plotted vertically. A gain of unity indicates the filter is passing all frequency components from 6.3 to 16 Hz. On either side of this bandwidth accelerometer signal components are attenuated or reduced at a rapid rate. The inverse of the standard is used to reflect the frequency sensitivity of human-body vibration, because the body does not respond to all vibration linearly. (These curves are analogous to the A-weighting curves of the ear used in noise.) The dashed lines represent the acceptable electronic tolerances of the filter.

In order to obtain a single number readout, we next electronically rectify the vibration signal and obtain its rms value, and the final step is to display this number on a meter. Thus the instrumentation arrangement shown in Fig. 4.29 allows one to obtain multiple vibration measurements simultaneously; to record same; to perform Fourier analysis sequentially from the tape recorder and finally to obtain a single number value for each of the vibration channels comparing these numbers to accepted standards. This arrangement allows for an optimal characterization of the vibration in the workplace.

The next question which arises is how does one obtain measurements with all of this electronic equipment in a workplace situation? For hand-arm vibration

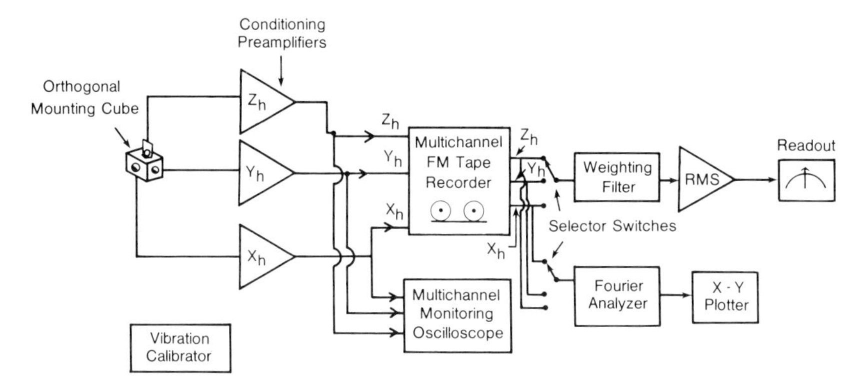

Fig. 4.29 Typical three direction (triaxial) vibration measurement system configuration.

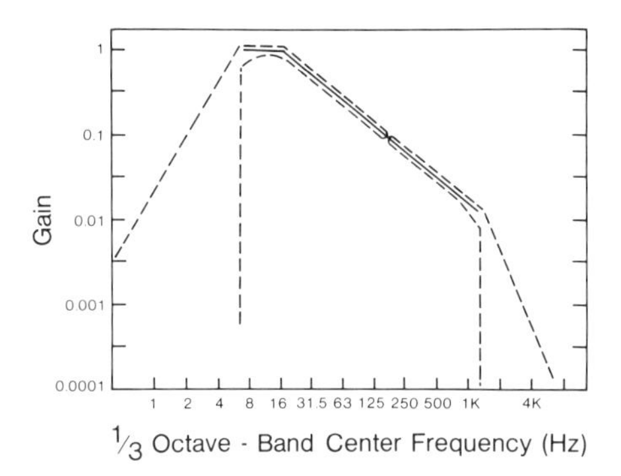

Fig. 4.30 Gain characteristics of the electronic filter network used to frequency weight acceleration components (continuous line). The provisional filter tolerances (dashed lines). (Courtesy ACGIH)

the situation is simple if a source of AC line voltage is available, or if need be a portable gasoline generator supplying AC can be used. For whole-body vibration the situation is similar in a plant site situation for hand-arm measurements. (The reader is encouraged to use an AC line regulator from the AC plant line to the vibration measuring equipment, because in many plant sites there are large electrical surges throughout the workday which the regulated power supplies in the vibration measuring equipment usually cannot fully accommodate. The use of an AC line regulator will smooth out these large pertubations and together with the regulated power supplies in the measuring equipment will usually permit measurements to be made over the workday). In the case of whole-body measurements obtained from <u>moving vehicles</u> there are two choices. The first is to securely mount the instrumentation package to the moving vehicle itself. However, the vehicle is vibrating while in motion and can potentially damage the instrumentation package unless it is carefully shock mounted to the vehicle. In this case separate batteries will need to be used. Our experience is to use a charged 12VDC automotive storage battery (not the vehicle's battery, because of possible electrical interference from the vehicle's electrical system). If all of the measuring equipment can be operated directly from the storage battery that's fine, but in many cases the 12VDC will need to power a solid state inverter which will produce sufficient AC line power to operate the measuring equipment. Using this method of using an instrumentation package on the moving vehicle, one cannot easily observe and monitor the vibration data as it is being obtained, (unless someone rides with the vehicle with an oscilloscope or chart recorder). We have thus chosen an alternate method: Radio frequency telemetry transmitted to a nearby receiving van (13,14) shown in Figs. 4.31a and b. The receiving van and its control center inside of the vehicle are shown in Figs. 4.32a and b.

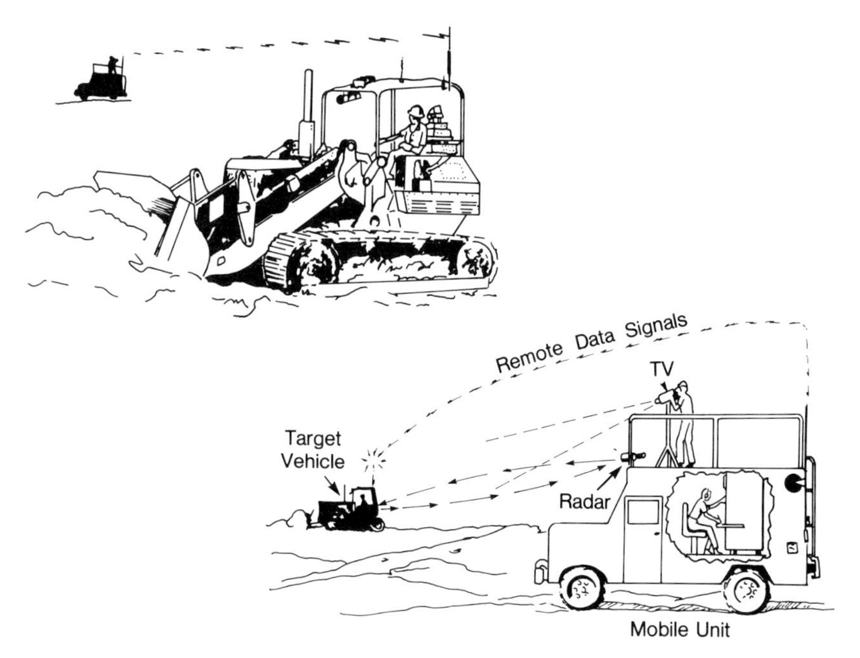

Fig. 4.31 The use of radio frequency FM telemetry to obtain whole-body vibration measurements. In (a) the operator is "wired" for transmitting measured data. In (b) the data is received, tape recorded, and analyzed by a nearby mobile unit. Vehicle speed is obtained using a Doppler radar unit and a synchronized videotape of the target vehicle and operator are also obtained. (Adapted from Wasserman et al, references (13,14))

A worker is "wired" into the moving vehicle, with a special telemetry package operating from 12VDC. This is an FM/FM telemetry system. Fourteen channels of vibration (and other environmental data) are obtained and frequency modulate fourteen subcarrier oscillators. These fourteen subcarrier channels are then mixed electronically. This composite signal then frequency modulates an R.F. carrier operating at 216.5 mHz. The transmitting distance is approximately 3-5 miles depending on the terrain conditions. There is also two-way R.F. voice communications between the vehicle operator and the receiving van command post. The transmitted data is received by the mobile unit. First the R.F. carrier is removed electronically, then each of the fourteen channels is simultaneously separated and FM tape recorded. Fourier analyzers, chart recorders, etc. are in the mobile unit and can monitor "on line" the quality of the incoming data. A video camera and VCR monitor the target vehicle's activities as the vibration data is received. The command post operator provides a narrative of the driver's activities which are simultaneously recorded both on the VCR and FM data tape. A police Doppler radar unit records the target vehicle's speed as it moves. This instrumentation arrangement provides the ultimate in obtaining vibration measurements in the workplace, and does not interfere with the worker. It is, however, an expensive investment in equipment. This example demonstrates how very sophisticated this technology can become depending on the measurement needs.

In the majority of cases for industrial hygiene and human factors purposes only a single number measurement is required, thus much of the sophisticated measurement instrumentation can be reduced as well as the cost. In the case of hand-arm measurements B&K has introduced a pair of hand-arm vibration meters shown in Fig. 4.33. One unit and its accelerometer are required per channel of vibration. The piezoelectric accelerometer is matched to the unit, thus no calibration is required. The accelerometer can be magnetically affixed to a vibrating hand tool, or to a hose clamp device previously shown in Fig. 4.20. The unit has the ISO5349 weighting characteristics built in, with a thermometer type readout giving either acceleration, velocity, or displacement. There is also an output jack for a recorder or FFT analyzer. One unit reads out in g's, the other unit reads out metrically in m/sec^2. These units are portable solid state battery operated devices.

For portable whole-body measurements, Endevco Corp., for example, has introduced a ride meter shown in Fig. 4.34. The ride meter is a single channel device consisting of a small electronics unit and is electronically weighted against Whole-Body Vibration Standard ISO2631 (to be discussed in a later chapter) with Liquid Crystal Display readout. The unit is portable and contains a rechargeable battery. The operator supports the meter with a small

Fig. 4.32 (a) External view of vibration mobile unit with roof mounted video camera and FM receiving antenna, (b) Interior view of instrumentation control center in vibration mobile unit. (Adapted from Wasserman et al, references (13,14))

attached harness. A single accelerometer is buried in the aforementioned hard rubber disc in the vertical direction. In use, as previously explained the disc is placed between the vehicle seat and the operator's buttock. There are provisions for attaching a recorder or an FFT analyzer to the unit.

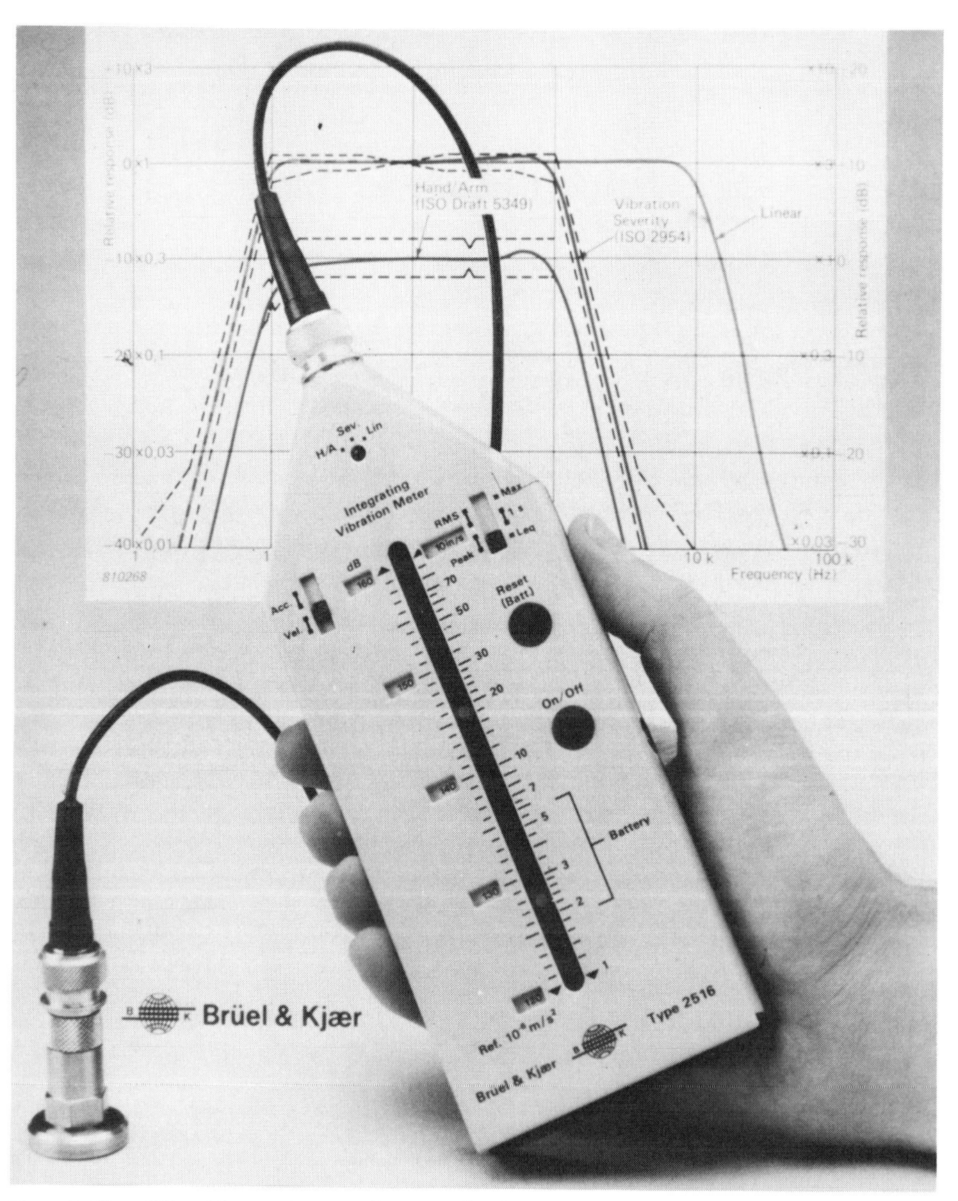

Fig. 4.33 Vibration meters and magnetically mounted accelerometer used for hand-arm measurements.
(Courtesy B & K Co.)

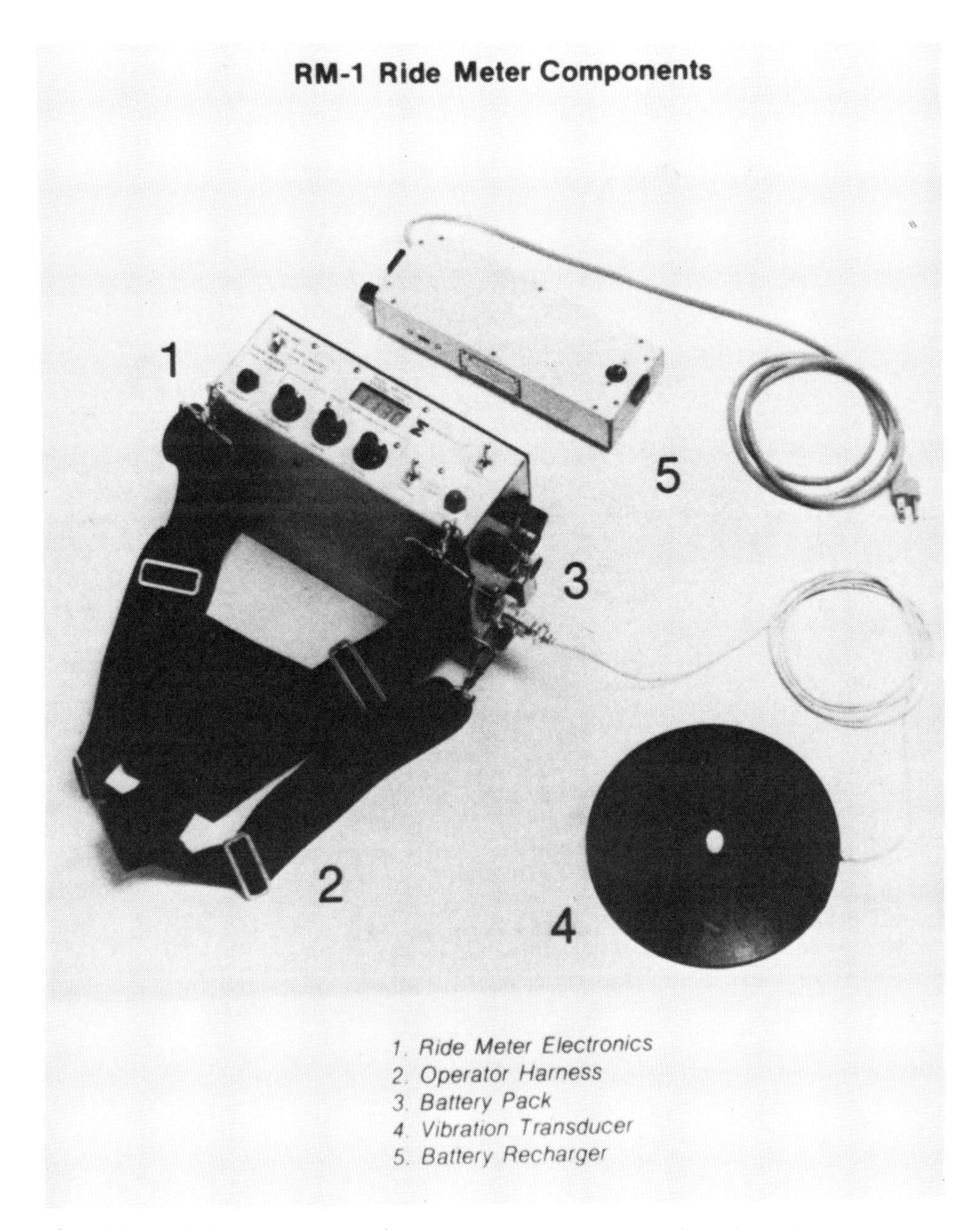

Fig. 4.34 Endevco portable single channel whole-body vibration ride meter (ISO 2631 weighted) with accelerometer disk and battery charger. (Courtesy Endevco Corp.)

When data is obtained and then FM recorded and can be analyzed leisurely, then the data reduction situation changes. B&K, for example, has introduced a very sophisticated combined hand-arm and whole-body vibration signal processor containing various weighting filters for hand-arm, whole-body, and motion

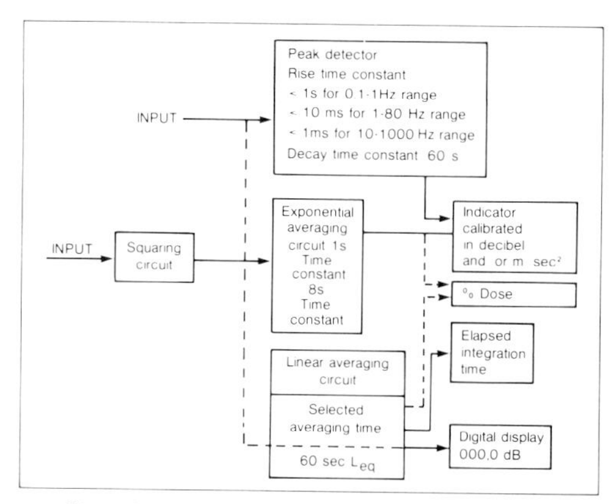

Signal processing and read-out choices

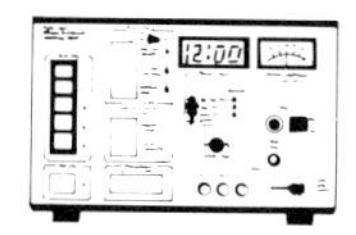

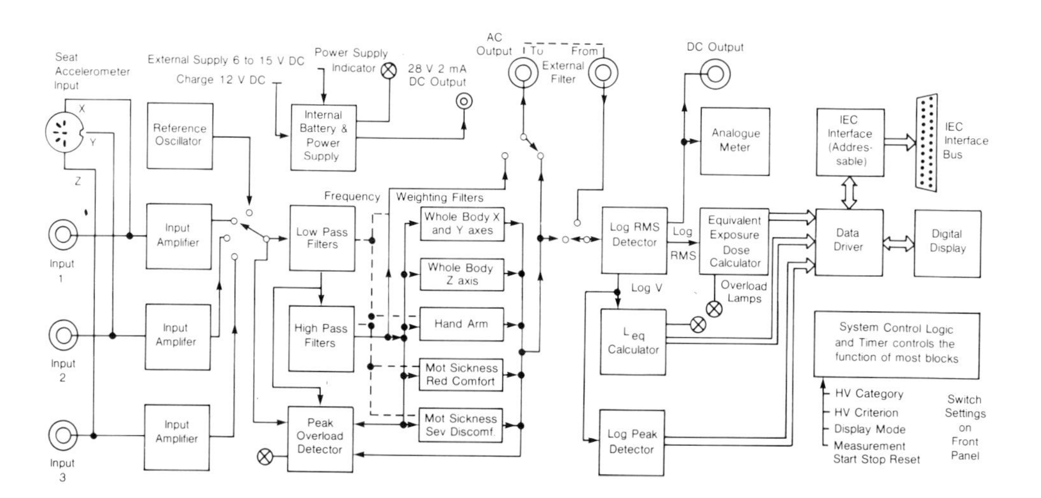

Simplified schematic block diagram of the Human-Response Vibration Meter

Fig. 4.35 B & K single channel combined whole-body (ISO 2631) and hand-arm (ISO 5349) laboratory human response vibration meter. (Courtesy B & K Co.)

sickness curves (shown in Fig. 4.35) and a myriad of operating and readout features. The devices' single disadvantage is that it will only accept one channel of vibration at a time. Thus each channel of tape recorded data would need to be played through the processor or multiple processors could be used simultaneously. We mention for completeness an undesirable measurement configuration which occasionally appears in the vibration literature, namely, the use of a portable sound level meter for measuring vibration. In order to do this, an accelerometer and integrator take the place of the sound level microphone which has been removed. The sound level meter is thus used as an rms measuring device. The problems arise with calibration of the system and the lack of proper instrumentation (e.g. sound level meter amplifiers are not charge amplifiers and have several electronics shortcomings) and it is therefore suggested that the use of a sound level meter for vibration measurements not be done (9). Some studies have been performed using proximity and optical motion devices. Their use is not widespread in human vibration measurements and its difficult to assess the accuracy of these techniques over conventional measurement techniques already described.

4.8 FOURIER SPECTRUM ANALYSIS (8,9)

In the previous chapter the theory of Fourier spectrum analysis was discussed. Now we would like to briefly discuss some of the more practical aspects of this analysis and what the frequency analyzer actually does.

The conventional vibration meter will give us a single vibration level measured over a wide frequency band. In order to reveal the individual frequency components making up the entire wideband signal we perform a frequency analysis (see Fig. 4.36).

For this purpose we use a computer with a filter which only passes those parts of the vibration signal which are contained in a narrow frequency band. The pass band of the filter is moved sequentially over the entire frequency range of interest so that we obtain a separate vibration level reading for each band.

The filter can consist of a number of individual, fixed-frequency filters which are frequency scanned sequentially by switching, or alternatively, continuous coverage of the frequency range can be achieved with a single electronically tunable filter.

There are two basic types of filter techniques used for the frequency analysis of vibration signals. The constant bandwidth type analysis where the filter has a constant absolute bandwidth, (for example 3 Hz wide, 10 Hz wide filter windows, etc.) or a constant percentage bandwidth analysis where the filter bandwidth is a constant percentage of the tuned center frequency, for example 3%, 10% etc. Fig. 4.36 shows graphically the difference in these two filter types as a function of frequency. Note that the constant percentage bandwidth filter appears to maintain a constant bandwidth, this is because it is plotted on a logarithmic frequency scale which is ideal where a wide frequency range is to be covered. On the other hand, if we show the two types of filter outputs on a linear frequency scale, it is the constant bandwidth filter which shows constant resolution. The constant percentage bandwidth filter plotted on a linear frequency scale shows an increasing bandwidth with increasing frequency which is not really practical.

Suffice to say there is no easy answer to the question of which type of frequency analysis to use. Constant percentage bandwidth analysis tends to match the natural response of many mechanical systems to forced vibrations, and allows a wide frequency range to be plotted on a compact chart. It is therefore the analysis method which is most generally used in vibration measurements. Constant bandwidth analysis gives better frequency resolution at high frequencies and when plotted on a linear frequency scale is particularly valuable for sorting out harmonic patterns, etc.

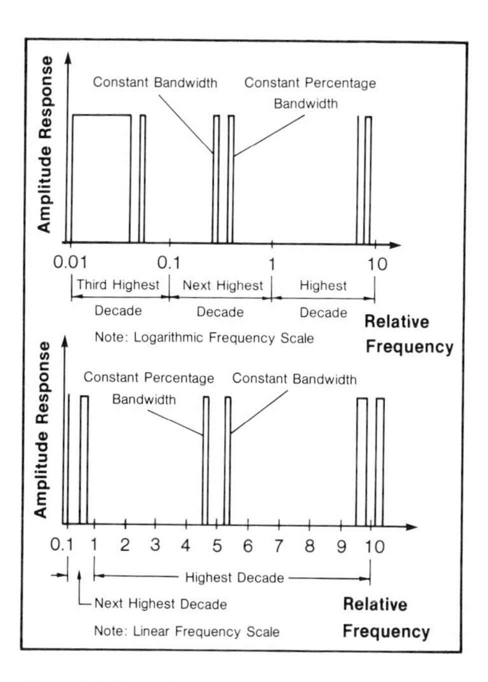

Fig. 4.36 Diagram showing two types of frequency analysis: constant percentage bandwidth and constant bandwidth in linear and logarithmic frequency scales (see text). (Courtesy B & K Co.)

4.9 CONDUCTING THE VIBRATION MEASUREMENTS STUDY IN THE WORKPLACE

4.9.1 Walk-Through Survey

Before attempting to perform any meaningful measurements it is advisable that one understand the role vibration plays in the work process. To merely walk into a work situation and begin making measurements is not advisable, rather one should take time to walk-through the entire work process and spend time observing how workers work, especially with regard to a vibrating process. Prior to conducting our own studies we used a portable video camera and VCR to actually record each step of the overall work process and then proceed to observe and record in detail the actual work process involving vibration. One usually discovers in these vibration processes that there is an actual repetition in work cycles which can be broken down into elemental steps. These elemental steps and the entire vibrating work cycle form the basis of the measurements which are to be made. Unless the vibrating process contains continuous unbroken vibration, it is advisable to observe and/or video tape several work cycles and actually time the entire work cycle and its various elements. Average time readings may be required, depending on the cycle time. Also, it is advisable to observe if possible several experienced workers as opposed to inexperienced workers performing their tasks. New workers on the job usually have had little training and have not had sufficient time to develop smooth and professional work habits.

The objective of measurements is to sample from a representative sampling of both experienced and inexperienced workers which reasonably represents what one would have achieved if all workers in the worksite were measured. To perform the latter would be impractical, expensive and would be mutually disruptive to the work process. Many employers are very hesitant to allow measurements which interrupt the work process for any length of time. Similarly, if the workers themselves are paid piece workers, they too are leery of taking work time out for measurements which may reduce their pay. In our studies, we found it sometimes advisable to actually use workers after their workshift to continue working while measurements were being made. This problem will vary depending on the attitude of the management, labor, and the worker. It cannot be overemphasized enough that observing the work process, subdividing the process, and careful planning are the keys to success to performing such studies.

4.9.2 Conducting the Workplace Study

After performing the walk-through survey, and if possible reviewing videotapes of the vibrating work process, one should be able to subdivide the work process into elements and elemental times. Before going further you must know what you seek from the data, is it simply to obtain a single dose value? If so, survey meters of the type described will be needed. Do you wish to analyze the data in detail? If so, an FM tape system, Fourier analysis, etc. will be needed. At how many different vibrating locations do you simultaneously need measurements? What type of instrumentation setup at each measurement location is needed? Will one or two or three accelerometers be needed at each of these locations? What equipment do you have? What equipment will be needed? Must you buy this equipment, or can you rent or lease some of the expensive equipment for a period of time (e.g. FFT analyzers)? What type of accelerometers and calibration equipment is necessary to perform the study?

If we assume you have answered these important questions, the next step is to obtain and completely familiarize yourself (and your team) with the measurement apparatus. One of the best ways to do this is to actually simulate a "dry run" of the measurements, recordings, etc. This can be done in the case of vibrating hand-tools by obtaining a tool and simulating work conditions outside of the main production area. In the case of whole-body plant measurements, try to use an unused vibrating machine, perhaps at off hours, etc. The idea is simply this: you should make all of your initial mistakes before you actually conduct the actual study with little or no interference with the actual work process, because what happens is that once the actual measurements are made in the workplace, interference with the work process must be absolutely minimal in order to avoid costly delays, disrupting production lines and angry management and workers.

Once you and your team are actually ready to perform the measurements, it is advisable to meet with the appropriate participants (e.g. workers, management, labor, IH's, etc.) and describe what it is you will do, how you will do it, how much disruption in the workplace there will be, etc. Finally, you are ready to perform the study. Our experience is that you must keep extensive logs of all activities including tape logs. We use a videocamera and videotape recorder observing the worker as we record the actual vibration data. One simple and inexpensive way of coordinating both the FM and videotape recordings is for a member of the team to orally narrate and log the recording activities by simultaneously speaking into two microphones, one connected to the VCR, the other connected to the voice channel of the FM tape system (10,13,14). When the data and videotapes are played back later for analysis, it is quite easy to synchronize both by matching the same voice tracks on both recorders. Little data and coordination is missed this way.

The reader should be aware that there are engineering groups world-wide, developing standardized and specific applications of vibration measurement techniques; for example, in Europe the PNEUROP group consists of various manufacturers of pneumatic tools and vacuum pumps. They are working on specific procedures for chipping hammers, and other pneumatic tools such as grinders, percussive tools, etc. In the U.S., the American National Standards Institute has similar committees. ISO has similar groups. (Several of these groups' addresses are given in the Supplementary Bibliography at the end of this chapter).

In summary, the material outlined in this chapter will allow the reader to perform most of the measurements encountered in the workplace. As appropriate, if your measurement problem is clearly unique and defined, it would be worthwhile obtaining and using specific standard or test procedure documents from the appropriate standards testing organizations.

Finally, some specific health and safety workplace standards per se will be discussed in depth in the next two chapters in relation to workplace measurements already discussed. These form the means of evaluating the data obtained in the workplace or other suitable standards can be used.

REFERENCES

1 International Standards Organization, Guide for the Measurement and the Evaluation of Human Exposure to Vibration Transmitted to the Hand, ISO5349, (TC108/SC4), Geneva, 1982.
2 International Standards Organization, Guide for the Evaluation of Human Exposure to Whole-Body Vibration, ISO2631, (TC108/SC4), Geneva, 1978. (Corresponds to ANSI S3.18-1979 and ASA38-1979)
3 W. Taylor, The Vibration Syndrome, Academic Press, London, 1974.
4 W. Taylor and P.L. Pelmear, Vibration White Finger in Industry, Academic Press, London 1995.
5 D.E. Wasserman and W. Taylor (Eds.), Proceedings of the International Occupational Hand-Arm Vibration Conference, DHEW/NIOSH, Public No. 77-170, 1977.
6 A.J. Brammer and W. Taylor (Eds.), Vibration Effects on the Hand and Arm in Industry, John Wiley and Sons Publishers, New York, 1982.
7 Columbia Research Corp. (Technical Staff) Shock and Vibration Measurements Handbook, 1971.
8 J.T. Broch, Mechanical Vibration and Shock Measurements, Bruel and Kjaer Corp., 1973.
9 Bruel and Kjaer Corp. (Technical Staff), Measuring Vibration, 1980.
10 D. Wasserman, D. Reynolds, V. Behrens, W. Taylor, S. Samueloff, and R. Basel, Vibration White Finger Disease in U.S. Workers Using Pneumatic Chipping and Grinding Hand-Tools - Vol. II - Engr. Testing, DHHS/NIOSH Public. No. 82-101, 1982.
11 R.R. Coermann, The Mechanical Impedance of the Human Body in Sitting and Standing Position at Low Frequencies, U.S. Air Force Aerospace Medical Research Labs. Report No. 61-492, 1961 (Ibid Human Factors, 4 (1962) 27-253).
12 J.C. Ziegart and J.L. Lewis, The Effect of Soft Tissue on Measurements of Vibrational Bone Motion by Skin-Mounted Accelerometers, Trans. Amer. Soc. Mech. Engrs., 101 (1979) 218-220.

13 D.E. Wasserman, T.E. Doyle, and W.A. Asburry, Whole-Body Vibration Exposure of Workers During Heavy Equipment Operation, DHEW NIOSH Public. No. 78-153, 1978.
14 D.E. Wasserman, W.A. Asburry, and T.E. Doyle, Whole-Body Vibration of Heavy Equipment Operators, Shock and Vibration Bulletin, 49 (1979) 47-68.
15 G. Rasmussen, Human Body Vibration Exposure and Its Measurement, Bruel and Kjaer Technical Review No. 1, 1982.
16 I. Pyykko, M. Farkkila, J. Toivanen, O. Korhonen and J. Hyvarinen, Transmission of Vibration in the Hand-Arm with Special Reference to Changes in Compression Force and Acceleration, Scan. J. Work Envir. and Health 2 (1976) 87-95.

SUPPLEMENTARY BIBLIOGRAPHY

Given below is a partial list of (nonmilitary) technical groups and organizations world-wide involved in and working on vibration testing procedures for the evaluation of specific whole-body, hand-arm, vibrating processes, tools, machines, vehicles, etc., commonly found in the workplace. Items in parenthesis indicate major standards areas of involvement.

Austria

Osterreichisches Arbeitsring fur Larmbekampfung, Regiersingsgebaue 1012 , Vienna (vib. machinery)

Osterreichisches Normunesinstitut(on), Meinstrasse 38, 1012 Vienna (Whole-body and hand-arm)

Bulgaria

Institat de Normalisation 8, rue Sveta Sofia, Sofia (vib. machinery)

Finland

Institutet for Arbetsbygien, Helsinki (hand-arm and whole-body)

France

L'Association Francaisse de Normaliatan (AFNOR) Tour Europe, 92 Courbevoic (vib. machinery)

Institut National de Recherche et de Securite, 30, rue Olivier-Noyer, 75680 Paris Cedex 14 (seats and pneumatic tools)

Germany

(DBR): Beuth Vertriebo GMBH, 1 Berlin 30, Burggrafenstrasse 4-7 und 5 Koln, Friesenplatz 167 (whole body and hand-arm)

(DDR): Amt fur Standardisiening D.D.R. Mohrenstrasse 37a, 108 Berlin (whole-body and hand-arm)

India

Indian Standards Institution, Manale Bhaven, 9 Bahadur Shah, Zafer Marg., New Delhi 1 (vib. machinery)

Japan

Japanese Institute of Standards, Hitotsugi-cho, Akasuka, Minato-Ku, Tokyo (vib. machinery, whole body and hand-arm)

National Institute of Industrial Health, 21-1, Nagao Gchome, Tamu-Ku, Kawasaki 213 (whole-body and Hand-arm)

Netherlands

Nederlands Normalisatie Instituut, Postbus 5059, 2600 GB Delft (vib. machinery)

Norway

Norges Standardiscringsferbund, Hakon 7, gt.z, Oslo 1 (vib. machinery)

Sweden

Svenska Elektriska Kommission, PO Box 5177, S 102-44-Stockholm, (whole body, hand-arm, vib. machinery)

Arbetarskyddsstysdsen, S-171-84 Solna (whole-body and hand-arm)

Forestry Occupational Health Services, Skolgartan 8, 92100 Lycksele (chain saws)

Switzerland

International Standards Organization (ISO), 1 Rue dc Varembe Geneva, Switzerland (whole-body, hand-arm, vib. machinery)

Swiss Gov't. Eidgenossiche Drucksaben und Material Zentrale, Bern 3 (whole body, hand-arm)

U.K.

PNEUROP (Technical Committee #17), British Compressed Air Society, 8 Leicester St. London, WCZH7BN (Procedures for measuring vibration from hand-held tools)

British Standards Institution (BSI), 2 Park St. London W.1 (whole body and hand-arm)

Food and Agriculture Organization/Economic Commission for Europe/ International Labor Organization Joint Committee on Forrest Working Techniques - Chain Saw Group on Hand-Arm Vibration. British Forestry Commission, Fornhaun, Surrey, U.K.(chain saw - hand-arm vibration)

Motor Industry Research Association, Watling St., Nuneaton, CV10-OTU (Whole body seat and vehicle dynamics)

Institute of Sound and Vibration Research - Univ. of Southampton, Southamptom, U.K. S09-5NH (whole-body and hand-arm vibration)

U.S.A.

American National Standards Institute (ANSI)/Acoustical Society of American - Stds. Secretariat, 335 East 45th Street, New York, NY 10017 (whole body and hand-arm)

Compressed Air and Gas Institute, 1230 Keith Building, Cleveland, Ohio 44115 (pneumatic hand-tools)

Society of Automotive Engineers, 2 Pennsylvania Plaza, New York, NY 10001 (whole-body vibration, vehicular seat measurements, and ride dynamics)

Amer. Foundrymen's Society, Golf and Wolf Roads, DesPlaines, Ill. 60016 (hand-arm, whole body in foundries)

Vibration Institute, 101 West 55th St., Clarendon Hills, Ill. 60514 (vib. machinery)

Society of Agricultural Engineers, P.O. Box 229, St. Joseph, Michigan 49085 (Agricultural vehicles and seats)

Amer. Conf. Gov't. Ind. Hygienists, 6500 Glenway Ave., Cinti., Ohio 45221 (hand-arm, whole-body)

Amer. Society of Mechanical Engineers, 345 East 47th St. New York, NY 10017 (machinery and vehicular vibration).

National Institute for Occupational Safety and Health (NIOSH), 4676 Columbia Parkway, Cincinnati, Ohio 45226 (Whole-body vibration).

U.S.S.R.

Office for Standards and Measurements, Prahal-Nove Mesto, Vaclauske Namesti 19, U.S.S.R. (whole body, hand-arm, vib. machinery)

Chapter 5

Hand-Arm Vibration Standards/Guides

5.1 INTRODUCTION

Previous chapters have examined the medical, epidemiological, and performance effects of both hand-arm and whole-body vibration. In addition basic vibration terminology, measurement transducers and instrumentation, and methods for performing workplace studies have been presented. In this and the following chapter, the major hand-arm and whole-body standards will be discussed as well as how they are used. One of the major problems with such workplace standards is that the user immediately assumes that such standards are the ultimate in worker protection, in most cases they are not. Thus, not only is the proper use of the standard important, but more important is the potential, unintentional, misuse of the standard. The scientists and engineers who develop these standards in working groups, as this author has learned through many years of participation on such groups, strive to present the best standards documents which the state of knowledge will permit, and then proceed to provide caveats, footnotes, appendices, etc. in the document in an effort to let the user know the limitations of the document's use. Unfortunately, many users either do not read these caveats or choose to ignore them, thus drawing misleading conclusions from their data. In this and the next chapter, not only will the salient points of the major standards be presented, but also in order to enlighten the reader some of the history of the standard as well as the advantages and criticisms of the standard will be presented with reference to its use in the workplace.

In this chapter only hand-arm standards will be presented. In the next chapter, whole-body standards will be presented. In all cases the reader is advised to obtain the complete standard before attempting to use the information in these chapters, since only salient parts of each document can be presented here.

There are many groups of national consensus standards organizations in the world, with very specific mandates. However, the two best known and largest of the international standards groups are the International Electrotechnical Commission (IEC) and the International Organization for Standardization (ISO). The former is concerned with electrical/electronics engineering issues, the latter is concerned with other issues such as vibration, noise and a variety of other issues. ISO consists of the national standards organizations of some 90 countries and lists over 5,000 published international standards. "ISO work is decentralized, being carried out by 163 technical committees and 649 sub-committees which are organized and supported by technical secretariats in 36 countries. The Central Secretariat in Geneva assists in coordinating ISO operations, administers voting and approval procedures, and publishes the International Standards. The people who develop

International Standards are an estimated 20,000 engineers, scientists and administrators. They are nominated by ISO member countries to participate in the committee meetings and to represent the consolidated views and interests of industry, government, labor and individual consumers in the standards development process. ISO coordinates the exchange of information on international and national standards, technical regulations and other standards-type documents, through an information network called ISONET which links the ISO Information Center in Geneva with similar national centers in 59 countries." (1). ISO, therefore, provides, as in the case of vibration, an international forum for individuals working in the field to bring together their collective knowledge to formulate consensus standards, which when published in its final form, became available for use. Most of the major standards to be discussed in these two chapters are ISO standards.

5.2 SOME MAJOR HAND-ARM STANDARDS/GUIDES

5.2.1 ISO/DIS 5349.2 (6)

The most widely known of the (international) hand-arm standards is ISO/DIS 5349.2 (Guidelines for the Measurement and Assessment of Human Exposure to Hand-Transmitted Vibration). At this writing, the standard is currently in the DIS or Draft International Standard status with full expectation that the current version of the document will very soon become an official ISO standard. This standard originated in 1969 in ISO Technical Committee 108, Subcommittee 4 from proposals at both the Japanese and Czeckoslovakian Vibration delegations to consider the effects of hand-transmitted vibration on humans (2). Most of the studies which comprised these original proposals were studies of subjective response and transmissibility of Miwa (3,4) and Louda (5) and NOT derived from hard VWF medical/epidemiological data. Figure 5.1 shows the basic "elbow" shaped curves adapted in the earlier versions of ISO5349. The horizontal axis of the graph is defined over a wide frequency band to accommodate the anticipated multiple tool spectra encountered in the workplace. The horizontal axis defines accelerations over 5 decades expressed in either m/sec^2 (rms) or g_{rms}, when 1g = 9.81 m/sec^2. The elbow shaped curves represent a family of frequency dependent curves for various exposure times during the workday. The curves have a horizontal (flat) portion over a low frequency range for constant acceleration in that range. As the frequency increases, the curves show acceleration changes and is proportaional to frequency (which represents constant velocity - see Chapter 3, equations 2-5). Acceleration values falling below each respective curve for a given exposure time are acceptable exposures. However, this does not necessarily mean that workers will not get VWF. It merely means that if values fall

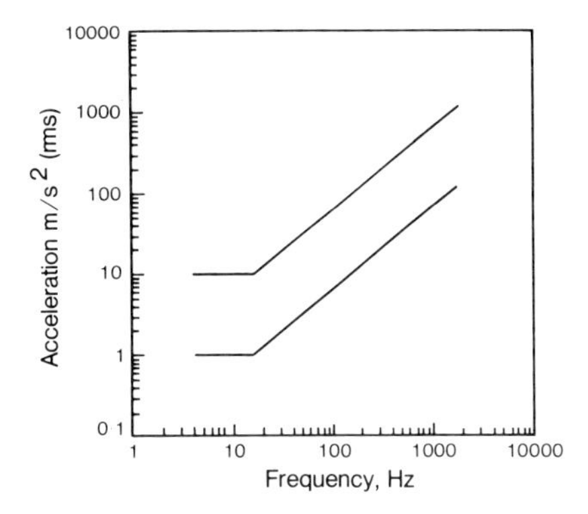

Fig. 5.1 Shape of the weighting function used in ISO 5349, DD43(1975), and ACGIH TLV standards as given in octive frequency bands versus vibration acceleration.
(From Griffin, reference (2) and ISO 5349, reference (6))

below their respective time dependency curves, workers might be safe based upon the state of knowledge at the time of issuance of the document. The reason is simply this, the majority of original studies which formed the basis of the document were not for the most part based upon hard medical/ epidemiological data. For example, in a given study just because a subject finds it subjectively intolerable while holding a vibrating handle and/or pressing his hand against a vibrating plate is no reason that he will get VWF. Thus, knowing some of the inherent weaknesses of the document the ISO working group of scientists charged with producing the document (of which this author is one) have through the years modified and reinforced the basic document with data obtained from actual VWF medical/epidemiological studies using the work of Brammer (7,8). This extensive revision of the basic document took place in 1981 at an ISO meeting in New Orleans and was further revised based upon salient comments with overall voting on this second revised document ending in July 1984. The new document is ISO 5349.2 and is discussed next. (For details, the entire document should be read carefully).

The standard consists of the body of the document and two annexes A and B. The user obtains triaxial measurements with respect to the biodynamic coordinate system shown before in Fig. 4.1 with an overall frequency range of 5 Hz to 1500 Hz (using either 1/3 octave bands centered from 6.3 to 1250 Hz or octave bands centered from 8 Hz to 1000 Hz). Vibration acceleration is in rms and is expressed in m/sec^2 or can also (as some have requested) be expressed in terms of decibels (dB) using 10^{-6} m/sec^2 as the reference according to the equation:

$$L = 20 \log(\frac{a}{a_r}) \quad (1)$$

where: L = acceleration level

a = measured acceleration in m/sec^2

a_r = reference acceleration, 10^{-6} m/sec^2

Acceleration should be measured for as long a period of time as possible using linear integration. In the case of short duration, accelerations from say percussive tools, extreme caution must be exercised in obtaining the signal and avoiding the pitfalls of D.C. shifting, destruction of acclerometers, etc (as described in Chapter 4). ISO recommends using a mechanical filter between the tool and the transducer whose bandpass (transfer) characteristics is linear to 3,000 Hz, thus minimizing the problems of low frequency attenuation of many mechanical filters (which could modify the actual vibration impinging on the accelerometers). Also advisable is a measure of the worker's grip force and hand orientation during the measurement, recognizing that both affect the amount of actual vibration reaching the hands; the harder the grip force, the more vibration reaching the hand; the lower the grip force, the less vibration reaching the hands. There are some who feel the document's overall frequency range should be extended much higher in frequency to 50 KHz (7).

As previously stated, what was needed was a measure of vibration dose versus human VWF response. It was not until 1981 when Brammer (8,9) performed an important study which provided the necessary information for the updated version of ISO 5349. He examined the data results of over 40 major epidemiological/medical studies which also included a vibration measurement component. These studies were of worker populations with exposures up to 25 years. He eliminated those studies where varied and multiple tool types and processes were used. He used populations where workers work year round and not in an intermittent manner. He used studies mainly which had defined the stages and the latencies used in the Taylor-Pelmear VWF classification system (10) (see Chapter 2 for details). Brammer then took the data from these studies, weighed the vibration measurements according to the 1979 revision of ISO 5349 using the "elbow shaped curves" and developed a series of curves of exposure time in years versus weighted acceleration (in either of the three orthogonal axis) for various percentiles of exposed populations for 4 hour/day exposures. Why 4 hour/day and not 8 hour/day? Because he found in most cases <u>actual</u> vibration exposure, for the studies he reviewed, normally did not exceed 4 hour/day. The VWF marker point he used was stage 1 of the Taylor-Pelmear system, namely, the appearance of the first white finger tip on exposed workers. This corresponds to the first appearance of peripheral vascular symptoms and signs, and determines the latent interval from which the worker was first exposed to vibrating hand tools to the appearance of the first white finger type. Figure 5.2 shows the Brammer findings and as used in the current ISO document. Thus for example, for 4 hour daily exposures of 50 m/sec²(or 5.10 g_{rms}) it will take 1.2 years if the worker is in the 50th percentile. However, if the vibration is reduced by a factor of ten to 5 m/sec² (or 0.5 g_{rms}), it will take this worker 14.1 years to reach stage 1. Similarly, if the worker is in the 20th percentile and has a 10 m/sec² vibration dose for a 4 hour day (equal to 1.02 g_{rms}), it will take about 4.1 years to reach stage 1. However, if the dose were reduced by half to 5 m/sec² (or 0.50 g_{rms}), he would reach stage 1 in a bit more than twice the time (8.5 years). However, if the 4/hr dose were reduced even further to 2 m/sec² (or 0.20 g_{rms}), than the time to stage 1 would jump up to 23 years! On the other hand, if the 4 hr dose went up to 50 m/sec² (or 5.10 g_{rms}), it would take only 0.7 years, or about 34 weeks before the appearance of his first white finger tip. Since the Brammer data has its limitations, these curves should not be used for exposure greater than 50 m/sec^2 or exposure times exceeding 25 years. It is also assumed that the worker has normal health, since complications from other disease processes can materially interfere with these predictions (see Chapter 2 and Table 2.1 for details). Also, the rate at which workers enter and leave their jobs affects these predictions; which leads to the next topic of intermittent daily exposures of hand-arm vibration.

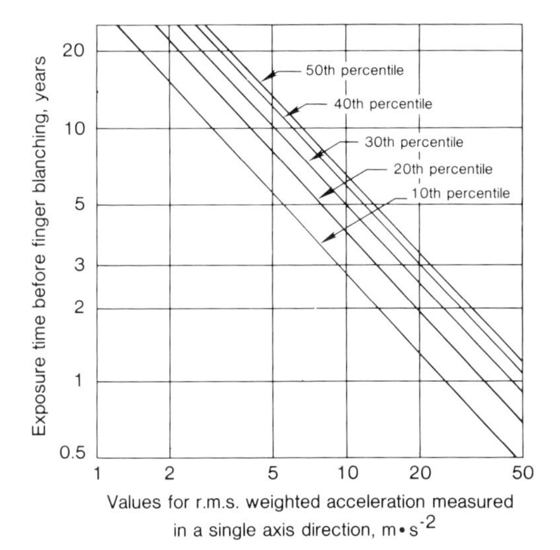

Fig. 5.2 Brammer curves for exposure times of percentiles of vibration exposed population groups in either X_h, Y_h, Z_h coordinate axis as used in and adapted by ISO 5349.
(From Brammer references (8 and 9))

If, for example, the acceleration of 20 m/sec^2 (or 2.04 g_{rms}) were measured for 2 hrs/day, then the equivalent acceleration level for the period of a 4 hour day can be determined by using the following equation:

$$A_{equiv} = \left(\frac{T}{4}\right)^{1/2} A_{meas} \qquad (2)$$

where a_{equiv} = equiv. accel. for a 4 hr. period

a_{meas} = meas. accel.

T = measurement period

Thus $A_{equiv} = \left(\frac{2}{4}\right)^{1/2} \times 20\ (0.7) \times 20 = 14.14$ m/sec² (or 1.4 g_{rms})

whereas if 20 m/sec occurred for the full four hours:

$$A_{equiv} = \left(\frac{4}{4}\right)^{1/2} \times 20 = 20 \text{ m/sec}^2 \text{ (or } 2.04\ g_{rms})$$

or if 20 m/sec^2 occurred for 8 hours:

$$A_{equiv} = \left(\frac{8}{4}\right)^{1/2} \times 20 = (1.4) \times 20 = 28 \text{ m/sec}^2 \text{(or } 2.85\ g_{rms})$$

If during the workday the total daily exposure consists of several exposures at <u>different frequency-weighted</u> acceleration levels, then the total frequency weighted acceleration can be obtained by using the following equation (3):

$$A_{equiv} = \left[\frac{1}{T} \sum_{i=1}^{n} (ak_i)^2 T_i\right]^{1/2} = \left[(ak_i)^2 \frac{T_1}{T} + (ak_2)^2 \frac{T2}{T} + \ldots (ak_n)^2 \frac{Tn}{T}\right]^{1/2} \qquad (3)$$

where $T = \sum_{i=1}^{n} Ti$

T = total daily exposure duration

ak_i = ith frequency-weighted, rms acceleration component with duration Ti

T_1 = that part of the daily exposure with rms acceleration ak_1

T_2 = that part of the daily exposure with rms acceleration ak_2

.

.

.

T_n = that part of the daily exposure with rms acceleration ak_n

As an example, let us assume that: $ak_1 = 12\ m/sec^2$ for $T_1 = 1$ hour

$ak_2 = 10\ m/sec^2$ for $T_2 = 5$ hours

then using equation (3) we get with $T = T1 + T2 = 1 + 5 = 6$ hours

$$A_{equiv} = \left[(12)^2 x \frac{1}{6} + (10)^2 x \frac{5}{6}\right]^{1/2} = \left[\frac{(144 + 500)}{6}\right]^{1/2} = 10.36\ m/sec^2$$

or (1.05 g_{rms}).

(Once again, the reader is encouraged to obtain the complete standard from the ISO at the address given under reference 6.)

5.2.2 ACGIH-hand-arm vibration TLV (11)

In the U.S. the American Conference of Governmental Industrial Hygienists (ACGIH) was organized in 1938 by a group of governmental industrial hygienists who desired a medium for the free exchange of ideas, experiences and the promotion of standards and techniques in industrial health. ACGIH is not an official U.S. Government agency. It is an organization devoted to the development of administrative and technical aspects of worker health protection. The association has contributed substantially to the development and improvement of official industrial health services to industry and labor. The committees on Threshold Limit Values (TLV) are recognized through the world for their expertise and contributions to industrial hygiene. There are more than 2600 ACGIH members world-wide.

Historically, with the conclusion of publication of the aforementioned NIOSH vibration studies indicating a high prevalence and short latent period of VWF in some U.S. foundries, and thus refuting the claim that VWF no longer existed in the U.S. (see Chapter 2 for details); the ACGIH Physical Agents TLV Committee began meeting with several scientists in the field in 1980. The Committee concluded that a hand-arm vibration TLV for the workplace needed to be developed for the U.S. since none existed. In 1982 the ACGIH (fully cognizant of the draft status of ISO/DIS 5349) asked some of these scientists for proposals of hand-arm TLV's with the important notion that a ACGIH TLV would be used in conjunction with a comprehensive package of other protective elements such as antivibration gloves, antivibration tools, work practices, etc. in order to remove VWF from the workplace (more fully discussed in chapter 8). The reason for this notion was and is that with the current state of tool technology, that a completely protective standard alone for hand-arm vibration most likely could not be met, but a comprehensive multielement package of protective measures would all contribute towards protecting the worker and could be practically used in the workplace. The TLV could be modified to become more protective with time, since the ACGIH bylaws permit TLV's to be periodically modified as new scientific and technological data become available. Thus the committee voted on and chose the hand-arm vibration TLV proposed jointly by Drs. W. Taylor, A.J. Brammer, and this author. It first appeared as a TLV in 1984.

In particular the TLV utilizes essentially the same coordinate system as ISO/DIS 5349 (previously shown in Fig. 4.2) with the ACGIH basicentric part originates on the cylindrical (2 cm) bar itself, where the accelerometer would most likely be actually mounted. Refer again to Figure 4.30, this shows the recommended characteristics of the ISO filter weighting network noting that the filter's bandpass is the reverse of the actual elbow-shaped curves from 5 Hz to 1500 Hz. The ACGIH TLV uses this same weighting filter.

Table 5.1 is the actual hand-arm vibration TLV for periods of up to 8 hours continuous exposure. The most important thing to be noted is that this TLV is protective from stage 3 upwards, not stage 1, of the Taylor-Pelmear classification system. Also a) Acute exposures for infrequent periods and times are not necessarily deemed more harmful, however, if these acute exposures extend to three times the TLV, and if enough occur, then it would be the same as 5-6 years of exposure. b) Accelerometers no greater than 15 g_{rms}

TABLE 5.1
ACGIH threshold limit values for exposure of the hand to vibration in either X , Y , Z directions.

Total Daily Exposure Duration*	Values of the Dominant,† Frequency-Weighted, rms, Component Acceleration Which Shall Not Be Exceeded a_K , ($a_{K_{eq}}$)	
	m/s^2	g
4 hours and less than 8	4	0.40
2 hours and less than 4	6	0.61
1 hour and less than 2	8	0.81
Less than 1 hour	12	1.22

*The total time vibration enters the hand per day, whether continuously or intermittently.
†Usually one axis of vibration is dominant over the remaining two axis.
If one or more vibration axis exceeds the Total Daily Exposure then the TLV has been exceeded.
(From reference 11 and Courtesy of ACGIH)

mass, with a cross axis sensitivity not exceeding 10% should be used for measurements. If necessary a mechanical filter whose upper frequently cutoff is 1500 Hz can be used to avoid the deshifting problems encountered in measurements.

For various intermittent daily vibration exposures equation 3 should be used to calculate the equivalent daily exposure.

5.2.3 British Standards Institute Draft for Development: DD43 (1975)

Griffin (2) carefully and succinctly traces the evolution of hand-arm standards in the U.K. prior to 1975 which led to the development of DD43 (1975): "In 1930 the Annual Report of H.M. Chief Inspector of Factories reviewed a study by Middleton (1930) of "dead hand" due to pounding-up machines in the boot and shoe industry. Far from calling for a vibration limit it is stated that "vibration, as such, is not an important factor, except in so far as it calls for an excess of static resistance in opposing the tendency to displacement of the object, and this depends on the amplitude of the hammer blows and not on their frequency."

A study by Hunter et al (1945) which was assisted by H.M. Inspector of Factories concluded that: "if the frequencies between 2000 (c.p.m.) and 3500 (c.p.m. (i.e., 33 Hz to 58 Hz) were avoided in constructing the tools, the incidence of white fingers would be very substantially reduced." Early researchers were aware of the operating rate of tools but did not measure the vibration frequency or level. Agate and Druett (1947) attempted to relate measurements of the vibration of portabled tools to the effects they produced. From a study of road rippers, pneumatic chisels, coal picks, rock drills, riveting hammers, scaling hammers and a portable grinding tool they say "it appears probable that vibrations of large amplitude between the frequencies of about 40 and 125 cycles per second are concerned in producing Raynaud's phenomenon, and that tools which show none below 600 cycles (per second) are not likely to produce it."

In 1967 and 1968 the Forestry Commission introduced a new process in which felling, debranching and crosscutting were all performed with chain saws. The increased exposure to vibration soon resulted in the high incidence of white finger which had been reported in other countries after large-scale use of the chain saw. In 1968 and 1969 the Forestry Commission adopted a vibration limit based on a proposal by Axelsson (1968) of Sweden. This required that the root-mean-square value of the amplitude of the vibration in the three axial directions should not exceed 80 microns (1 micron = 10^{-6} meters) in the frequency range 50 to 500 Hz. (Saws in common use in 1968 are reported to have had values of 300-600 microns as determined by this procedure.) After

1969 the Forestry Commission used the first Czechoslovakian proposal to the International Standards Ornization. This was interpreted as providing a limit of 18 m/s^2 rms in the 63 Hz octave band and 32 m/s^2 rms in the 125 Hz (and higher frequency) octave bands. The interest in vibration limits within the Forestry Commission was a strong influence on the subsequent activities of the British Standards Institution.

The form of the "Guides to the evaluation of exposure of the human hand-arm system to vibration" is partly based on a proposal to the International Standards Organization. The frequency weighting is based on research conduced in Japan by Miwa (1967) in which he determined equivalent comfort contours for vibration of the hand over the frequency range 3 to 300 Hz. The panel of the BSI adopted the shape of a simplified approximation to Miwa's curve (constant acceleration below 16 Hz and constant velocity above 16 Hz. However, the frequency range was extrapolated to 4 to 2000 Hz.

A maximum weighting of 10 for variations in exposure time had been proposed in Czechoslovakia. This was retained by the British Standards Institution. However, the vibration levels and durations associated with these weightings were changed. Most obviously the changes sought to accommodate the known work pattern, vibration levels and incidence of Vibration White Finger in the Forestry Commission. This was the origin of limits for 150 and 400 minutes vibration exposure per day. The vibration levels selected for these two curves resulted in early saws exceeding the upper limit while later saws with anti-vibration mounts fell between the two limits (see Figure 5.1). The chain saw vibration data were effectively restricted to one frequency (125 Hz) but the apparent reduction in the incidence of new cases of white finger due to the adoption of new anti-vibration chain saws gave some credibility to the two curves. Early drafts employed a complex time dependency but this was later deleted for simplicity and only two limits were provided. These were conveniently locked to 1.0 m/s^2 rms over the range 4 to 16 Hz. (The 'simplification' of the guide to two limits was intended to assist its use and emphasize its approximate nature. However, as published, the labelling of the two limits may be confusing).

The two limits proposed in the BSI DD43 are shown in Figure 5.1. Analysis is to be conducted in octave bands and it is implied (though not clearly) that the limits should be evaluated separately for each octave and separately in each of three orthogonal axes of the hand."

Griffin then goes on to carefully note what he believes are the shortcomings of the standard: "There is a brief mention of the methods of mounting accelerometers and analyzing the vibration signals but the method will not ensure either accurate or repeatable results, especially with intermittent and time varying vibration. Perhaps the most obvious problem with the Draft for Development is the labelling of the two limits and the allowance for exposure time. Section 7 of DD43 states: "it is recommended that cumulative exposure time to vibration should never exceed 400 minutes; for intermediate periods between 150 minutes and 400 minutes some interpolation should be made; the values given for 150 minutes apply for all shorter periods. These latter values should not be exceeded by regular users of hand-held equipment." So, for regular use the upper curve should never be exceeded and exposures should not be longer than 400 minutes. It appears that reducing the exposure from 6 hours 40 minutes to 2 hours 30 minutes will allow an increase of 10 in the vibration level. This is very different from the Czechoslovakian standard and ISO proposals on which it was based. These allow no change in level for duration greater than 2 hours but a tenfold change from 30 minutes to 2 hours. (Some insight into the manner in which the above differences arose may be partly inferred from a paper by Keighley (1974) published before the Draft for Development had been finalized. He indicated that the upper curve was to apply when there is continuous exposure of less than half-an-hour or regularly interrupted exposure totally not more than 2 1/2 hours per working day. Keighley also states that the limits were not to apply to continuous exposures of less than half-an-hour or interrupted exposures totalling 2 1/2 hours. These should be regarded as occasional usage and not subjected to any limit unless there is evidence of VWF. This is different from the published Draft for Development which indicates that the upper curve applies to cumulative exposures of 150 minutes.)

In its final form the Draft for Development gives insufficient guidance on the interpolation method by which to calculate limits for periods between 2 1/2 hours and 6 hours 40 minutes. Additional problems concern the definition of exposure duration, how to evaluate exposures which change in level from moment to moment, and how to sum exposures of various levels and durations.

Referring to the Draft for Development it has been stated that "levels above the 150 minutes line with exposures of more than 20 minutes continuously per day or more than sixty minutes spread over the day are also likely to cause VWF." Keighley, (1975). This contains an implied contradiction to the statement in the Draft for Development that levels above the 150 minute limit should not be exceeded by regular users of hand-held equipment.

Irrespective of whether the shape of the frequency weighting given in B.S.I. DD43 is appropriate to the prevention of Vibration White Finger, it is clear that there are problems associated with the interpretation of the Draft for Development. It has provided guidance where little existed and served as a useful basis for comparing vibration measurements. However, it does not provide a complete procedure for assessing many real complex vibration exposures." (2)

5.2.4 Swedish hand-arm vibration standards (13)

For many years the Scandanavian countries have contributed much to the knowledge of VWF and vibration syndrome (which they call TVD or Traumatic Vasospastic Disease). Particularly, the epidemiology and laboratory work of Drs. Ing-Marie Lidstrom, J.E. Mansson, S. Axelsson, and colleagues in Sweden, and I. Pyykko and colleagues in Finland have for many years contributed to both understanding and eliminating VWF from the workplace (12-14).

When the gasoline powered chain saw was introduced in Sweden in the early 1950's, no one was really aware of the VWF problem. However, awareness came quickly with the appearance of high prevalences of VWF. When the National Swedish Testing Institute for Agricultural Machinery measured vibration in chain saws in 1974, the average vibration level of the chain saws used in Sweden had been reduced to 30 to 40 Newtons force from some 150 Newtons in 1967 as a result of manufacturer's technical improvements.

Since 1973, the National Swedish Board of Occupational Safety and Health has restricted the maximum permissible vibration level of chain saws sold in Sweden to 50 Newtons, where 50 Newtons force corresponds to 90-100 m/sec^2 or 9.2-10.2 g_{rms} (See Fig. 5.3). The National Swedish Testing Institute for Agricultural Machinery began in 1965 to develop a technique for measuring vibration of chain saws. Based on the limited experience at that time, a threshold limit value expressed in amplitude displacement over frequency was proposed: 0.08 mm in the 50 to 500 Hz frequency range.

The report (15) of the preliminary vibration measurements of available chain saws stated:

> "With regard to vibration, practically all the chain saws now on the Swedish market have a vibration level in excess of the proposed injury limit. The difference in maximum amplitude between different saws is very slight. Professional use of present-day chain saws in modern logging operations where the tempo is rapid, the daily utilization time long and recuperation breaks few and far between, together with the regular use of the chain saw for limbing as well, would appear to involve appreciable risk of the eventual onset of vibration injury (vasoconstriction) in the fingers. Moreover, the vibrations impose a physiological load on the human frame which tends to impair, and in many cases heavily impairs, work performance (see Table 5.2)."

TABLE 5.2
Swedish threshold values of physiological load through effects of vibration on hands and arms during work with a power saw.

Vibration force in Newtons	Subjective effects	Effects on work performance
10	Noticeable, not uncomfortable	No impairment
10-30	Noticeable, but scarcely uncomfortable	No impairment
30-50	Quite noticeable, prolonged exposure uncomfortable but tolerable	Very slight impairment
50-100	Distinctly noticeable, unpleasant after an hour or so; still tolerable	Performance impaired but work still possible
100-150	Uncomfortable, barely tolerable after more than 10 minutes' exposure	Performance heavily impaired; more than 10 minutes's work without a break inadvisable
150-300	Even brief exposure highly uncomfortable	Work difficult to perform
over 300	Extremely uncomfortable	Work impossible

(From reference 15)

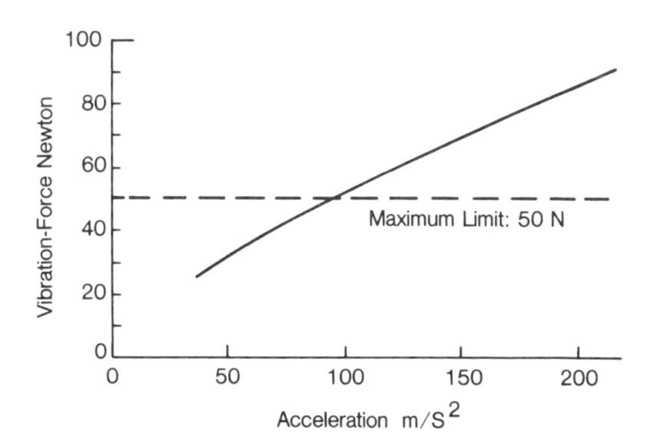

Fig. 5.3 Approximate relationship between acceleration and vibration force as used by the National Swedish Board of Occupational Safety and Health.
(From Axelsson, reference (13))

Faced with the serious results of the medical investigation in 1967--49% VWF among chain saw operators, the National Swedish Board of Occupational Safety and Health decided to regulate the vibration levels of chain saws. Although no fully accepted threshold limit values for vibrations existed at that time, the National Testing Institute had developed methods for measuring vibrations in chain saws. In 1971, the Board of Occupational Safety and Health issued the regulation that chain saws sold in Sweden should be vibration tested by the National Testing Institute. The maximum permissible vibration level of the saw handles was restricted to 80 Newtons in 1971, 60 in 1972, and 50 in 1973. This Swedish maximum permissible vibration level of 50 Newtons is not a scientific threshold limit value. It is more of a "qualified guess," decided in a serious situation. This "qualified guess"has, however, proved quite realistic with the passage of time (13).

TABLE 5.3

Frequency weighted rms hand-arm vibration levels used by the Swedish Board of Occupational Safety and Health over a frequency range of 6.3-1250 Hz.

Exposure duration during typical workday (hours)	Acceleration X, Y and Z directions	
	m/sec^2	grms
4	2.9	0.30
2	4.1	0.42
1	5.8	0.59
0.5	8.3	0.85

(From reference 16)

5.2.5 Czeckoslovakian, Russian, and other East European standards (2,18)

The first Czeckoslovakian hygienic standards appears to be heavily based on ISO/DIS 5349, which is not surprising since Dr. Louda's Czechoslovakian data was heavily used in the document's early development and he served on and later was the chairman of the ISO working group developing the document.

According to Griffin (2) in his review states that the U.S.S.R originally published its 191-55 Hygiene Standard in 1955, which was later superseded in 1966 by Sanitary Standard 626-66 and then complemented in 1972 by hand-tool standard GOST 17770-72(23), see Figs. 5.4, 5.5 and 5.6.

TABLE 5.4

Correction factors for various exposure durations as given in the Czechoslovakian Hygiene Regulation No 33 (1967).

(Corrections under A and B are alternative and may not be used together.)

A. UNINTERRUPTED OR IRREGULARLY INTERRUPTED VIBRATION EXPOSURES:

Vibration exposure during 8 hour shift	Correction in dB
up to 30 minutes	20
30 minutes to 1 hour	10
1 hour to 2 hours	5
more than 2 hours	0

B. REGULARLY INTERRUPTED VIBRATION EXPOSURES:

Duration of interval without vibration	Number of interruptions of vibration during shift	Correction in dB
Up to 2 minutes	0	0
2 to 10 minutes	less than 5	0
	5 to 10	5
	more than 10	10
More than 10 minutes	1 or 2	5
	2 to 5	10
	5 to 10	15
	more than 10	20

(As cited by Giffin - reference 2)

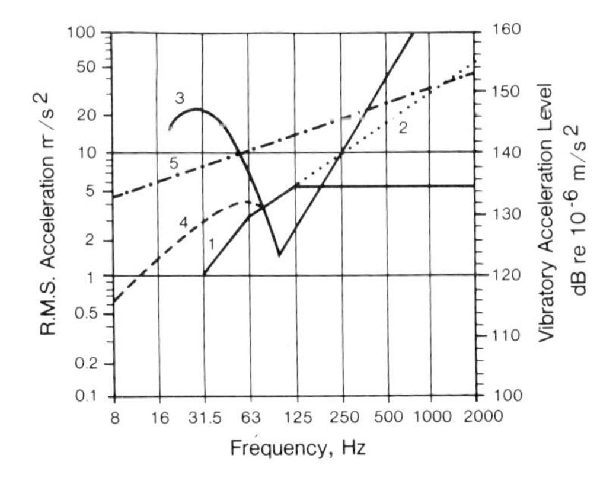

Fig. 5.4 Various Czechoslovakian and USSR hand-arm vibration standards (see text).
(From Louda and Lukas, reference (18))

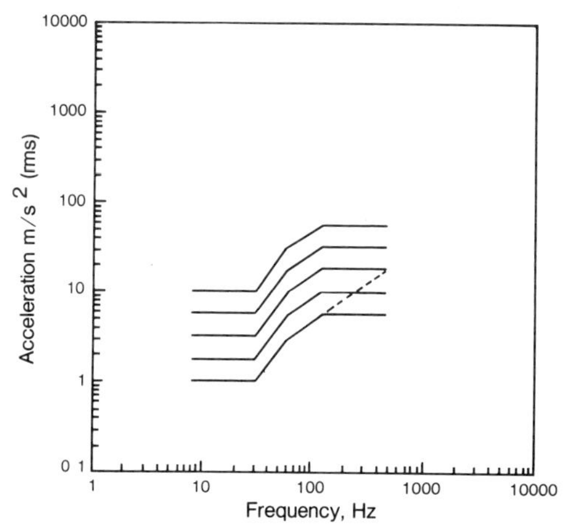

Fig. 5.5 Octave band vibration limits defined in Czechoslovakian Hygiene Regulation No. 33 (1967). Broken line defines Louda's proposal above 125 Hz (see text and curve 1, Fig. 5.4 and correction table 5.4).
(Adapted from Griffin, reference (2) and Louda and Lukas, reference (18))

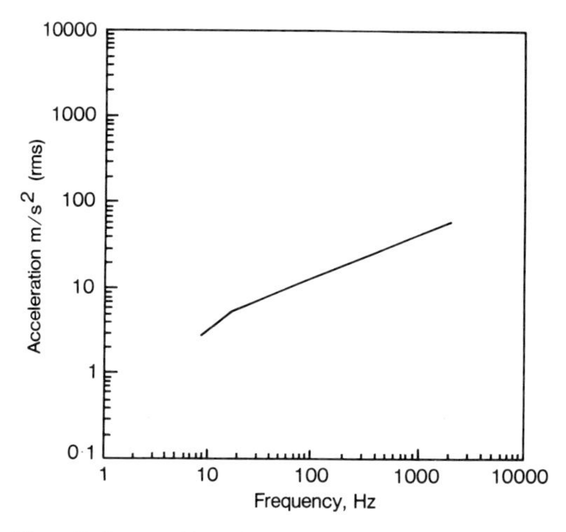

Fig. 5.6 Vibration limit for hand-tools defined in USSR Standards 626-66 (1966) and GOST 17770-72 (1972).
(From Griffin, reference (2))

In summary, some of the major hand-arm vibration standards have been briefly presented, as well as some of the lesser known developmental steps and difficulties which have led to these documents. The intent is to sensitize the reader that each of these documents have both advantages and disadvantages and that these standards should not be used blindly or else incorrect conclusions can be drawn. It is imperative that the complete standard be obtained and carefully studied and reviewed for proper use. The reader is referred to the Supplementary Bibliography (Chapter 4) for additional names and addresses of groups involved in standards work for additional information.

REFERENCES

1. ISO in Brief-Information Bulletin, ISO, Geneva, Switzerland, Feb. 1985.
2. M.S. Griffin, Vibration Injuries of the Hand and Arm: Their Occurrence and the Evolution of Standards and Limits, Institute of Sound and Vibration Research, Univ. of Southampton, Technical Report No. 101, July, 1979.
3. T. Miwa, Evaluation Methods for Vibration Effects - Measurements of Unpleasant and Tolerance Limit Levels for Sinusoidal Vibrations, Ind. Health (Japan), 6 (1968) 18-27.
4. T. Miwa, Evaluation Methods for Vibraion Effects - Measurements of Threshold and Equal Sensation Contours on the Hand for Vertical and Horizontal Sinusoidal Vibrations, Ind. Health (Japan), 5 (1967) 213-220.
5. L. Louda, Mechanisms Prevosu Vibraci Prumysloveko Zdroje na Cloveka. Kandidatsker Disertacni Prace, CVUT Fakulta Stronjniho Inzenyrstvi Praha, 1965.
6. ISO/DIS 5349.2, Guidelines for the Measurement and Assessment of Human Exposure to Hand-Transmitted Vibration, International Organization for Standardization, Case postale 56, CH-1211, Geneva 20, Switzerland, 1984.
7. R. Dandanell and K. Engstrom, Vibrations From Percussive Tools Such as Riveting Tools in the Frequency Range 6 Hz to 10 mHz and Raynaud's Phenomenon, Proc. 4th. International Symposium on hand-Arm Vibration, Helsinki, Finland, May, 1985.
8. A.J. Brammer, Threshold Limit for Hand-Arm Vibration Exposure Throughout the Workday, in Vibration Effects on the Hand and Arm in Industry, A.J. Brammer and W. Taylor, (Eds.), Wiley and Sons Publishers, New York, 1982.
9. A.J. Brammer, Towards Standards for Occupational Exposure of the Hand to Vibration, Proc. of the International Workshop on Research Methods in Human Motion and Vibration Studies, New Orleans, Sept. 1981 (In Press).
10. W. Taylor and P. Pelmear (Eds.), Vibration White Finger in Industry, Wiley and Sons Publishers, New York, 1982.
11. American Conference of Government Industrial Hygienists, Threshold Limit Values for Chemical Substances and Physical Agents in the Work Environment, ACGIH, 6500 Glenway Ave, Cincinnati, Ohio 45211, 1985.
12. I.M. Lidstrom, Vibration Injury in Rock Drillers, Chiselers, and Grinders, in Proc. International Occupational Hand-Arm Vibration Conference, D. Wasserman and W. Taylor (Eds.), DHEW/NIOSH Public. No. 77-170, 1977.
13. S.A. Axelsson, Progress in Solving the Problem of Hand-Arm Vibration for Chainsaw Operators in Sweden, 1967 to Date (Ibid.)
14. I. Pyykko, J. Hzvarinen, and M. Farkkila, Studies on the Etiological Mechanism of the Vasospastic Components of Vibration Syndrome, in Vibration Effects on the Hand and Arm in Industry, A.J. Brammer and W. Taylor (Eds.), Wiley and Sons Publishers, New York, 1982.
15. S.A. Axelsson, Analysis of vibrations in power saws. Studia Forestalia Suecica, No. 59, Royal College of Forestry, Stockholm, 1968.
16. J.E. Hansson, An Ergonomic Checklist for Industrial Trucks, National Swedish Board of Occupational Safety and Health Report No. 25, 1983.
17. A. Olsson, SS-ISO/DIS 5349 Vibration Och Stot-Riktlinjer for Matring Och Bedonning Av Handoverforder Vibrationer, SEK, PO Box 5177, 10244, Stockholm (Jan. 1986).

18. L. Louda and E. Lukas, Hygienic Aspects of Occupational Hand-Arm Vibration, in Proc. International Occupational Hand-Arm Vibration Conference, D. Wasserman and W. Taylor (Eds.), DHEW/NIOSH Public. No. 77-170, 1977.
19. Vremennye sanitarnye pravila i normy po ograniczeniyu vliyania vibracii na rabotayushczich rucznym pnevmaticzeskim i electriczeskim intrumentom v proizvodstve. USSR Ministry of Health Nr. 191-55, Moscow 1955.
20. E.C. Andreeva-Galanina and M.N. Belikov, Gigieniczeskaya i techniczwskaya charakteristika novych klepalnych molotkov i podderzek s umenshennymi otdaczey i vibraciey. Teoria i praktika gigieniczeskich isseldovaniy. LSGMI, 1985.
21. B. Kryze, K otazce hygienickeho normovani vibraci. Pracovni Lekarstvi 14 (1962) 198-194.
22. Smernice o ochrane zdravi pred nepriznivym pusobenim mechanickeho kmitani a chveni (vibraci). Czechoslovakian Ministry of Health 29, No. 33, 1967.
23. Mashiny rucznye. Dopustimye urovni vibraciy. GOST 17770. Moscow, 1972.

Chapter 6

Whole-Body Vibration Standards/Guides

6.1 INTRODUCTION

This chapter's focus is on whole-body standards/guides. Unlike hand-arm vibration where there is a definable biological endpoint (e.g., vibration syndrome), it would appear that whole-body has no such counterpart. There is to this day continued research and debate about the biological/physiological effects of whole-body vibration exposure. In the area of very low frequency vibration (0.1-1 Hz) motion sickness results under very specific conditions (see Chapter 2). Thus because of the above, the major standards/guides for whole-body vibration are directed more towards subjective discomfort and the ability to perform tasks rather than on health.

As in the previous chapter, some historical information relating to the standards will be given as well as criticisms of these standards. The reader should obtain the complete standards document and thoroughly understand its use and its limitations before attempting to use it in the workplace. The information contained herein should aid in that understanding.

6.1.1 ISO 2631 (1)

The most widely known and major whole-body document is ISO 2631 (Guide for the Evaluation of Human Exposure to Whole-Body Vibration). This document's history began in 1964 with ISOTC 108 vibration experts (many of the same group who later developed the hand-arm document ISO 5349) met. The first draft of this document appeared in 1967 and went through a series of revisions. It was issued as an international standard in 1974 with 19 of 21 members voting affirmative (with the UK and USSR voting negative) and later reissued in 1978 with minor changes (2). At this writing, there is again an extensive revision underway to update the document by the Committee.

Vibration acceleration measurements are obtained with reference to a standardized biodynamic coordinate system (see Fig. 4.1) where a_x represents "back to chest" motion, a_y represents "right-to-left side" motion and a_z represents ventral "foot (or buttocks) to head" motion. Only linear acceleration is used in the document. The exposure curves for each axis is shown in Fig. 6.1 (a_z) and Fig. 6.2 (a_x and a_y). These curves assume both a frequency dependence (1-80 Hz) and a time dependence (1 minute to 24 hours). What we have in both these figures is a family of same shaped curves extending from 1 minute to 24 hours as a function of acceleration (in rms, m/sec^2 or g's). Fig. 6.1 depicts the U-shaped curves, where the base of each curve shows constant and lowest acceleration in the 4-8 Hz resonance band with acceleration increasing on both sides of the 4-8 Hz resonance band. Similarly in Fig. 6.2 when resonance for the a_x and a_y directions are in the 1-2 Hz band acceleration increases with frequency.

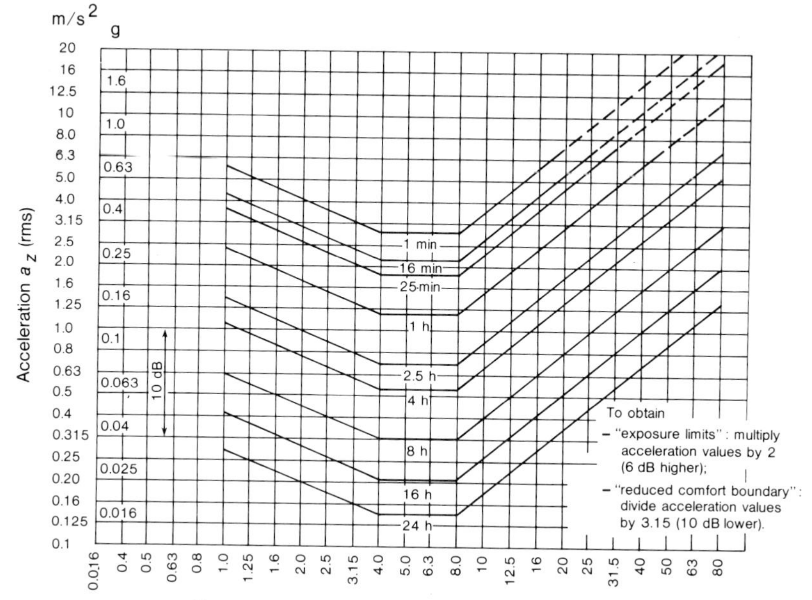

Fig. 6.1 a_z acceleration limits as a function of vibration frequency and exposure time. The curves are for Fatigue Decreased Proficiency boundary (FDP).
(From reference 1 - ISO 2631-1978)

Although Figs. 6.2 and 6.3 show a single family of curves there are actually three sets of curves for each figure. In particular the curves shown are referred to as "Fatigue Decreased Proficiency" boundaries (FDP) and represent the ability of a person to work at task(s) under vibration exposure without the vibration interfering with their ability to perform the work. The second set of curves is called "Reduced Comfort" (RC) boundaries and are concerned with preservation of comfort during the work process, riding in a vehicle, etc. This family of curves looks the same as the FDP curves except all acceleration levels are divided by 3.15 and are thus 10 dB lower than the FDP curves shown. The final set of curves is referred to as "Exposure Limits" and an attempt at preserving health and safety of persons under vibration exposure. These curves once again are the same shape as the FDP and RC curves but are 6 dB higher than the given FDP curves. Multiply the FDP accelerations by two and one obtains the exposure limits. (Also note that accelerations on rms values given in meters/sec^2 and g's, when 1 g = 9.81 m/sec^2 and frequency bands are given in third octave bands.) There are also special weighting correction factors given in the document depending on the axis of vibration. When vibration consists of a single discrete (sinusoidal)

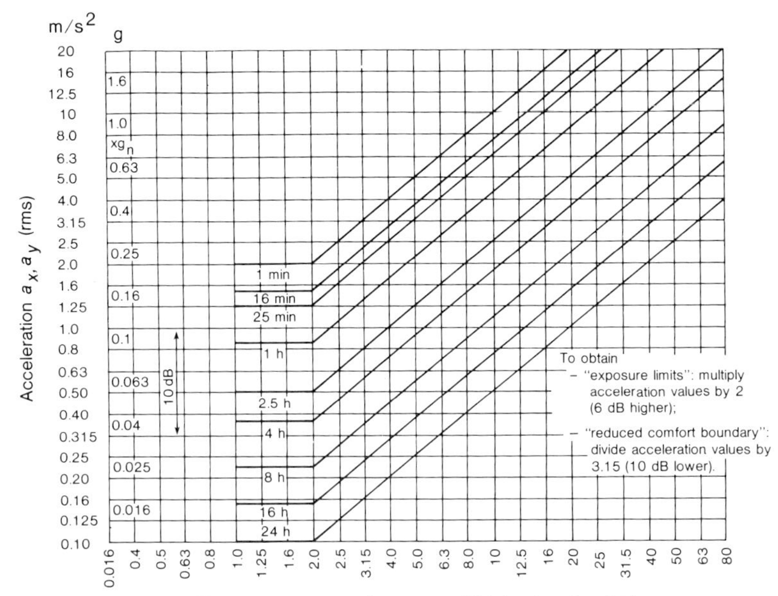

Fig. 6.2 a_x and a_y acceleration limits as a function of vibration frequency and exposure time. The curves are for Fatigue Decreased Proficiency boundary (FDP). (From reference 1 - ISO 2631-1978)

frequency, one consults Figs. 6.1 and 6.2 for the given condition to see whether or not the desirable limit (e.g., FDP, RC, or EL) has been exceeded for the respective exposure time. However, if multiple frequencies are present, then Fourier spectrum analysis is necessary. Then each frequency acceleration value comprising the measured spectrum must be compared to Figs. 6.1 and 6.2 to determine if the accelerations have exceeded one or more of the three limits. Similarly for random (narrow band vibration) and broad band vibration, each frequency component must be examined with respect to the appropriate limit at the center frequency of the respective band.

In the case of multiple direction vibration, each direction is examined separately as before. The vector sum of the vibration appearing in each of the perpendicular directions usually is calculated as the square root of the sum of the squares of the total vibration appearing in each axis. However, there has been a recent amendment to the basic document which recommends that "if two or three vectorial component at a multiaxis vibration have similar magnitudes when the a_x and a_y components are multiplied by 1.4, since the effect on comfort and performance of the combined motion can be greater than that of a single component" (3). Thus what ISO suggests is to first frequency weight each of the axis (using a filter network which is the reverse of the exposure curves given either in Fig. 6.1 or 6.2 or both). Next obtain the appropriate a_x, a_y, a_z values. Finally use a modification of the vector sum method given in the equation below:

$$a_t = \left[(1.4a_x)^2 + (1.4a_y)^2 + a_z^2\right]^{1/2} \tag{1}$$

where a_t = total equivalent acceleration

A second amendment is currently being drafted at this writing and is concerned with extending the low frequency limit from 1 Hz down to 0.1 Hz to account for ship motion vibration levels.

There is also the matter of high peak acceleration values where crest factors are used (defined as the ratio of peak acceleration to rms acceleration). This standard claims calculation validity for crest factors up to ratios of 6 with a minimum evaluation period of one minute, above 6 the standard underestimates the effects of the motion.

Since the basic ISO 2631 curves use a time dependency, continuous exposure is obtained from the respective curve values and corrections. If the daily exposure is interrupted but the acceleration values remain constant, then the effective total daily exposure is the sum of the individual exposures. If the daily exposure is interrupted but the acceleration values are considerably different, then the standard recommends a rather obtuse method for determining equivalent exposure times. The standard itself should be consulted in making these calculations. Finally, the 2631 curves are to be used for people of "normal health," which is undefined in the document.

In the U.S. the American National Standards Institute and the Standards Secretariat, Acoustical Society of America have issued the ISO 2631 standard as ANSI(S3.18-1979) and ASA38-1979, respectively.

Since there is a paucity of hand medical/epidemiological data for whole-body vibration and since ISO 2631 is one of the very few such documents available, some investigators have criticized some of its major shortcomings (2,4,5) in an effort to minimize the misuse of the document by unwary users and to reinforce the need for periodic revisions as new data becomes available; the latter is being a normal practice of ISO. In particular some of the major cited difficulties with the document are as follows: a) where a mixture (e.g., spectrum) of vibration frequencies is present in, for example, a work situation, then the documents limits are not necessarily a good predictor of performance. It would appear that the reasons for this, in part, is due to the short term discrete sinusoidal laboratory data from which much of the basic ISO 2631 curves were derived. b) The curves should only be used with people of normal health, but in reality workers exposed to vibration are not necessarily of normal health and suffer diseases and maladies just like other people. This restriction may be due to the fact that many of the studies which formed the 2631 curves were performed using young physically fit military personnel. c) Most of the data for the ISO curves were derived for the vertical direction, whereas the lateral vibration curves were derived on data from a small number of subjects (10-12) from Miwa (6) and Dieckmann (7). d) Probably one of the most controversial issues of whole-body vibration exposure is whether or not there is a "time dependency" associated with FDP, RC, and EL. Work by Miwa (6) and Simic (8) suggests there is, however, Oborne (2) in his critical review of ISO 2631 states the following:

> "This paper has critically assessed the current International Standard for human exposure to whole-body vibration (ISO 2631). The conclusion must be reached that the standard lacks empirical support in a number of important areas, although it is supported in others. In particular, the concept of time dependency for aspects such as performance and comfort has been shown to have no experimental basis, and the use of the same shape of frequency weightings for the three criteria of EL, FDP, and RC is an oversimplification of the true situation. Furthermore, the experimental bases upon which a number of other aspects included in the standard are founded have been shown to be either poor or nonexistent. Nevertheless, the shape of the frequency weighting for vertical vibration at least, and the weighting method suggested for nonsinusoidal vibrations are supported by subsequent data."

Because of some of the above concerns with ISO 2631, NIOSH and the nearby U.S. Air Force Aerospace medical Research Labs (Dr. H. VonGierke) agreed to cosponsor a study (4) to be conducted by Dr. J. G. Guignard and associates then at the University of Dayton Research Institute. The purpose of the study was to actually evaluate the FDP curves with sinusoidal inputs at recommended levels using three experimental regimens: a) 2 to 4 Hz using 16 to 25 minute short term exposures; b) 2 to 4 Hz using 2.5 to 8 hours long term exposures; c) 8 to 16 Hz using 16 to 25 minute exposures. A battery of 14 tests of peripheral and central performance as well as heart rate and urine analysis were performed on a variety of healthy subjects. The authors concluded from the study that "vibration at the ISO FDP levels was found to have negligible effect on the performance or physiological state of subjects when compared with the control runs."

The question remains, how useful is ISO 2631 in predicting subjective and performance effects in the actual workplace situation? As a typical example the work of J.E. Hansson and B.A. Wikstrom at the National Board of Occupational Safety and Health in Sweden is cited (10). The purpose of their study was to compare physical vibration measurements at the buttocks of the operators of several off-road forestry machines and compare each driver's subjective evaluation of discomfort to the recorded vibration levels in a variety of different driving situations. Vibration evaluations and measurements were made according to ISO 2631. The conditions and results were as follows:

> "A total of 13 different vibration analysis methods were considered. The investigation consisted of seven studies. The machines were driven on five test tracks, each consisting of six to ten shorter intervals representing easy to moderately difficult terrain conditions. Subjective ratings correlated better with technical evaluations based on the two most dominant vibration directions or all three directions than with only the critical direction according to ISO 2631. Calculations based on vibration energy in the entire frequency range 1-80 Hz gave better correlation than calculations based on energy in the critical frequency band according to ISO 2631. The weighted sum of vector method gave the best correlation with subjective ratings. The correlation coefficient for this method was clearly higher than with the method recommended in ISO 2631, i.e., 1-3-octave band critical direction and frequency" (10).

6.1.2 Other whole-body vibration and related standards

Since ISO 2631 represents the basic and most widely known of the whole-body documents, some other standards and guides have evolved from this document. For example, ANSI 3.29-1983 (same as Standard ASA48-1983)-Guide to the Evaluation of Human Exposure to Vibration in Buildings is a specialized application of ISO 2631 to people inside various buildings, based on acceleration and velocity perceptions of persons working in these buildings together with accompanying mathematical factors attempting to take into account the acceptability of a given magnitude of vibration in these buildings from 1-80 Hz. Similarly, there is another document which goes down to very low frequencies (0.063 to 1 Hz) for fixed and offshore structures (ISO 6897-1984, Guidelines for the Evaluation of the Response of Occupants of Fixed Structures, Especially Buildings, and Offshore Structures to Low Frequency

Horizontal Motion) to determine the subjective ability of persons to function in these structures.

There are several specific product improvement related standards affecting the workplace as well, for example, ISO 7096-1982 (Earth-Moving Machinery-Operator Seat-Transmitted Vibration); ISO/TR 5007-1980 (Agricultural Wheeled Tractors-Operator Seat-Measurement of Transmitted Vibration and Seat Dimensions); ISO 5008-1970 (Agricultural Wheeled Tractors and Field Machinery - Measurement of Whole-Body Vibration of the Operator); ISO-DP 8002 (Mechanical Vibration - Land Vehicles - Reporting Measured Data); ISO/DP 8041 (Human Response Vibration Measuring Instrumentation).

There are also a few documents which are concerned with human measurements other than pure acceleration, for example, ISO/DIS 7962 (Vibration and Shock - Mechanical Transmissibility of the Human Body in the (Vertical) Direction) and ISO 5982-1981 (Vibration and Shock - Mechanical Driving Point Impedance of the Human Body). Finally, in order to keep terminology uniform, there are two documents of interest ISO 2041-1975 (Vibration and Shock Vocabulary) and ISO 5805-1981 (Mechanical Vibration and Shock Affecting Men - Vocabulary).

Undoubtedly, there are other documents issued by individual countries which cover specifics of areas of local interest.

6.1.3 The absorbed power concept for whole-body vibration (11,12)

In the early 1960's the U.S. Army Tank-Automotive Center in Warren, Michigan began a classic program in attempting to forecast human response to vehicle vibration in military vehicles. The concept is called "absorbed power" and was introduced by Pradko and Lee (11,12) and later extolled by Janeway (13) as the criteria to be adapted in lieu of ISO 2631. In particular Pradko and Lee proposed that a person's response to vibration can be determined by measuring only the input vibration conditions. They proposed a single number result without prior knowledge of the frequency spectrum at the vibration impinging on the person. Pradko and Lee subjected persons to vibration and measured the input and output motions and the phase relationship between the two. They also measured the "absorbed power" or rate of energy absorbed by the person's body. They claimed other correlations with subjective responses of these subjects and that their response to vibration is "a function of the absorbed power and consequently that a constraint absorbed power corresponds to a uniform degree of subjective response" (13). Since power is a scalar and not a vector quantity, therefore they claimed that the vibration effects can be measured directly by summing each directional components contribution to the overall power. Their absorbed power in watts is given by:

$$P = K(A)^2 \quad (2)$$

when k = a constant for a given frequency and direction of motion

A = acceleration (rms) in feet/sec^2

Pradko and Lee next generated several "k" value tables for various accelerations and go on to demonstrate that their proposed conceptually is valid (11,12).

Unfortunately, the absorbed power concept was never accepted by the scientific community and was short lived.

REFERENCES

1. ISO 2631, Guide for the Evaluation of Human Exposure to Whole-Body Vibration, International Organization for Standardization, Case Postule 56, CH-1211, Geneva 20, Switzerland, 1978.
2. D.J. Oborne, Whole-Body Vibration and International Standard ISO 2631: A Critique, Human Factors 25 (1983) 55-69.
3. ISO 2631-1978/Amendment 1, Guide for the Evaluation of Human Exposure to Whole-Body Vibration, International Organization for Standardization, Geneva, Switzerland, 1982.
4. L. Louda, Perception and Effect of the Mixture of Two Vertical Sinusoidal Vibrations on Sitting Man, Work-Environment-Health 7 (1970) 62-66.
5. H.H. Cohen, D. Wasserman, R. Hornung, Human Performance and Transmissibility Under Sinusoidal and Mixed Vertical Vibration, Ergonomics 20 (1977) 207-216.
6. T. Miwa, Y. Yonekawa, and A. Kajima-Sudo, Measurement and Evaluation of Environmental Vibrations, Part 3, Vibration Exposure Criteria, Industrial Health (Japan) (1973) 185-196.
7. D. Dieckmann, A Study of the Influence of Vibration on Man, Ergonomics, 1 (1958) 347-355.
8. D. Simic, Contribution to the Optimumization of the Oscillating Properties of a Vehicle: Physiological Foundations of Comfort During Oscillations, Technical University of Berlin Dissertation D38, 1970 (Translation 1707-RAE, Feb. 1974).
9. J.C. Guignard, C.J. Landrum, and R.E. Reardon, Experimental Evaluation of International Standard (ISO 2631) for Whole-Body Vibration Exposure, University of Dayton Research Institute, Tech. Report No. UDRI-TR-76-79, 1976.
10. J.E. Hansson and B.O. Wikstrom, Comparison of Some Technical Methods for the Evaluation of Whole-Body Vibration, Ergonomics 24 (1981) 953-963.
11. F. Pradko, R.A. Lee and J.D. Green, Human Vibration-Response Theory, Amer. Soc. of Mech.s Engrs. Report No. 65-WA/MUF-19, 1965.
12. F. Pradko, R. Lee and V. Kaluza, Theory of Human Vibration Response, Amer. Soc. of Mech. Engrs. Report No. 66-WA/BHF-15, 1966.
13. R.N. Janeway, Human Vibration Tolerance Criteria and Applications to Ride Evaluation, Amer. Soc. of Automotive Engrs. Report No. 750166, 1975.

SUPPLEMENTARY BIBLIOGRAPHY

R.W. Shoenberger, Subjective Effects of Combined Axis Vibration: II. Comparison of X-axis and X-plus-pitch Vibrations, Aviat. Space, Envir. Med. 56 (1985) 559-563.

M. Demic, Definition of Recommendations for Assessment of Random Vibration Lateral and Longitudinal Movements of Seats in Transport Vehicles, Doc. No. ISO/TC 108/SC4/WG2N197, 1985.

M. Demic, Physiological Attitude to Definition of Tolerable Levels of Random Verticle Vibration Loads of Bus Underbodies from the Aspect of Comfort, Doc. No. ISO/TC108/SC4/WG2N/27, 1984.

L.M. Cleon, Measurement and Analysis of the Vibrations to Which Passengers and Staff are Exposed When Travelling in Railway Vehicles, Doc. No. ISO/T408/SC4/WG2N149, 1985.

D.J. Oborne, A Critical Assessment of Studies Relating Whole-Body Vibration to Passenger Comfort, Ergonomics 19 (1976) 757-774.

Chapter 7

The Control and Elimination of Vibration in the Workplace

7.1 INTRODUCTION

In this chapter, the question we want to answer is how to control and/or eliminate harmful vibration from the workplace, and protect the worker from its effects. Just because we have knowledge that there is a detrimental agent in the workplace, does not necessarily mean that it is easily eliminated. On the contrary, there are many factors to be taken into account and merely declaring there is a given standard which can be used in the workplace, in many cases is just inadequate. What needs to be determined are the realities of the workplace; the state of technology as to eliminating the agent from the workplace; and the realization that change in attitudes of workers, management, labor, etc. come painfully slow, with or without a given workplace standard.

7.2 THE ELIMINATION AND CONTROL OF HAND-ARM VIBRATION IN THE WORKPLACE

The control and elimination of VWF from the workplace is multifaceted, especially with regard to pneumatic tools where the vibration levels and consequent VWF is high. At this writing, it would be unrealistic to expect that antivibration (A/V) tool designs themselves can solve the problem because of the current state of the technology art. Also A/V gloves alone cannot solve the problem, neither can a standard alone or even work practices alone. Thus it would appear that a complete "package" which includes A/V tools and gloves, standards, work practices, and a sense of awareness and education in the workplace together would be the best combination of elements to solve this problem. Thus, ALL of the following are recommended (1-3):

7.2.1 Antivibration tools

Since the early 1970's gasoline chain saw manufacturers have been aware of VWF and have designed a series of A/V chain saws (4). Where these saws have been extensively used in place of standard nonA/V saws, there has been a falling incidence of VWF (4,5). Today nearly all gasoline chain saw manufacturers have at least one A/V saw, if not more, in their product lines. These A/V saws are available to those who want and need them. Acceleration levels have fallen from the 20-30 g range by a factor of ten or more to the 1-3 g range. A tribute to excellent engineering. These saws must be maintained and their shock absorbers periodically replaced, etc. in order to keep vibration levels low.

Unfortunately, this same success does not exist in most pneumatic tools. Only a very few pneumatic tool companies (principally Atlas-Copco and ESAB in Sweden and Vast-Hardill Co. in the U.S.) have chosen to design A/V pneumatic tools (see Figs. 7.1-7.4, the ESAB scaling hammer is distributed in the U.S.

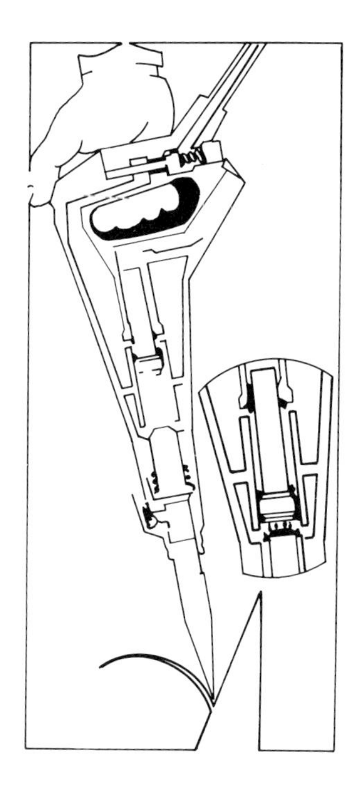

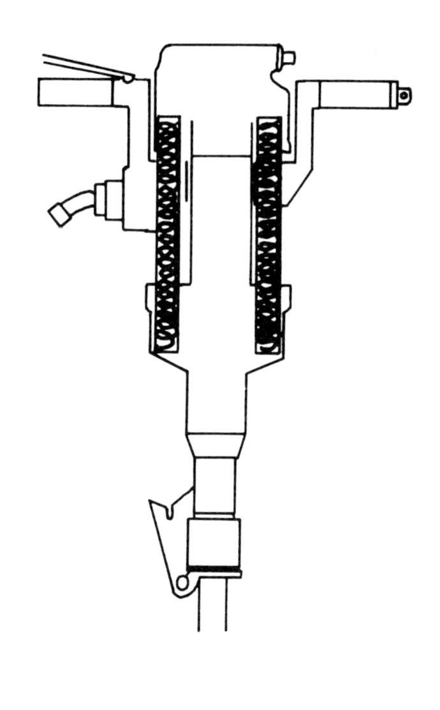

Fig. 7.1 Cutaway view of the Atlas-Copco antivibration pneumatic chipping hammer.
(Courtesy Atlas-Copco Co.)

Fig. 7.2 Cutaway view of the Atlas-Copco antivibration pneumatic pavement breaker.
(Courtesy Atlas-Copco Co.)

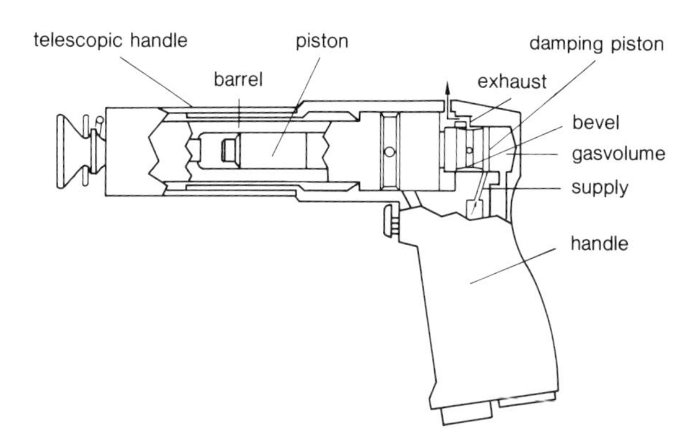

Fig. 7.3 Cutaway view of the Atlas-Copco vibration damped pneumatic riveting gun.
(Courtesy Atlas-Copco Co.)

by the ARO Corp. of Bryant, Ohio). In most cases the original pneumatic tool designs go back to the early 1900's (4-7).

Through the years several ingenious isolation and shock absorbing modifications to standard tools have been tried and for the most part have been found to be only marginally successful, these include placing protective sleeves over chisels, and handles, internally dampening chisels, dipping the handles of pneumatic tools into liquid shock absorbing materials which later harden, etc. (8,9). It would appear that several A/V modifications are currently being used in some east European countries, but their details are not widely known.

Finally, because these pneumatic A/V designs are new, their long term effectiveness for reducing VWF are not known at this writing.

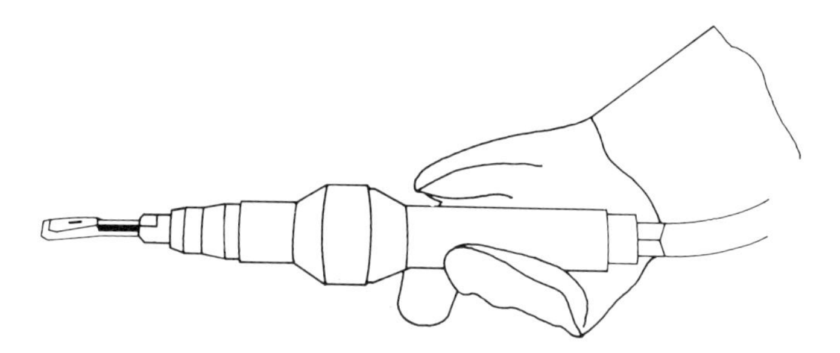

Fig. 7.4 The ESAB-AB/ARO antivibration pneumatic scaling hammer. (Courtesy ARO Corp.)

7.2.2 Antivibration gloves

The next line of defense against VWF is personal protection for the worker, namely, antivibration gloves. At this writing, there are four manufacturers of A/V gloves in the U.S., they are:

a) Shock Stop gloves made by Wolverine Co., P.O. Box 8735, Grand Rapids, Michigan 49508.

b) Sorbothane gloves made by the Sager Glove Co., 60 E. Palatine Road, Prospect Heights, Illinois 60070.

c) Guard-Line Inc, P.O. Box 919, Atlanta, Texas 75551.

d) Steel Grip Safety Apparel Co., Danville, Illinois 61832.

The Wolverine gloves used a patented material called PORON; the Sager gloves use a patented material called Sorbothane (originally developed for runners shoes). These two materials are known to be effective in removing lower frequency vibration in addition high frequency vibration and yet retain their overall shape.

What about the effectiveness of these materials? In a recent study by Goel and Rim (10) at University of Iowa in Iowa City, Iowa, a study was performed determining the effects of energy absorbing gloves in reducing the vibration experienced by the operator chipping grey iron castings under simulated laboratory conditions. The resulting accelerations experienced during the chipping were recorded on a four channel FM tape recorder using a set of accelerometers mounted on the back of the operator's hand used for holding/ guiding the chisel. The data were digitized and 1/3 octave band rms acceleration levels from 6.3 Hz to 4 kHz were obtained. The results indicated that the use of antivibration gloves did reduce the vibration on the hand holding the chisel significantly in comparison to the bare hand. The average values of the percentage reductions achieved for the materials tested leather, Poron padded, and Sorbothane padded gloves were 41%, 46% and 67%, respectively. These results are very encouraging. Thus, workers are encouraged to begin using A/V gloves. However, once again since these gloves are so new that one cannot speculate at this time as to their long term effectiveness in reducing VWF in the workplace.

7.2.3 Hand-arm vibration standards

These standards have been extensively addressed previously and will not be repeated here. They currently include ISO 5349.2, ACGIH-TLV, and related standards.

7.2.4 Suggested hand-arm vibration work practices (1-3)

Control and elimination of VWF from the workplace is multifaceted and all of the following are recommended as a work practices guide: a) Workers should, where possible, use both A/V tools and A/V gloves; b) TLV's and standards should be appropriately and carefully applied; c) Determining vibration exposure times and introduce work breaks (e.g., 10 minutes per continuous exposure hour) to avoid constant vibration exposure; d) Where possible, measure and monitor the vibration acceleration of tools to avoid increased acceleration with increased tool usage; e) Include a specialized replacement medical exam (11) for all workers especially those with a previous history of peripheral vascular and neurological abnormalities, and Primary Raynaud's Disease; f) Workers are advised to have several pairs of warm A/V gloves and warm clothing. Do not allow the hands to become chilled; g) Reduce smoking while using vibrating handtools to avoid the vasoconstriction effects of nicotine; h) Let the tool do the work by grasping it as lightly as possible, consistent with safe work practice, rest the tool on a support or workplace as much as possible. Operate the tool only when necessary to minimize exposure and at reduced speeds if possible; i) If symptoms of tingling, numbness, or signs of white or blue fingers appear, workers should be promptly examined by a physician.

7.2.5 Education and problem awareness

To many people, change in the workplace does not come easily. The reasons are many and varied: loss of job; loss of pay; there is no VWF here so why change a process; costs of protecting the worker; fear of lawsuits; dislike of government regulations, etc. Some of the reluctance to change can be altered through education of the workers, labor, management, physician, nurses, industrial hygienists, etc. Training courses; written materials (such as books of this type); proper use of and maintenance of vibrating tools; videotapes such as NIOSH Videotape #177 - Vibration Syndrome (2,12); posters, etc. all play an important part in heightening the sense of awareness of the VWF problem. Through this knowledge comes a realistic attitude about what can and cannot be done in the workplace. Customizing and implementing the elements in this package approach will demonstrate how VWF can be removed from the workplace.

7.2.6 Other protection

There are some other measures in addition to the above which in specific instances could help eliminate VWF from the workplace. In some cases the workplace can be redesigned especially when stationary (stand) grinders are concerned, when swing grinders might be able to take their place. In instances where the handtool not be handled at all (e.g., robotics) or where the worker merely guides the tool to do its work without actually holding and therefore coupling to the tool; these represent alternatives to conventional working. In the case of incentive (piece) workers, NIOSH found (13) that VWF prevalences were higher with shorter latent periods with incentive chipper and grinder workers studied than with other chipper and grinder workers (in the same foundries) who did not work under the incentive system. There needs to be a constant tool maintenance plan to keep A/V tools and conventional tools in first class condition (e.g., no dull chisels, properly maintained workings, periodic replacement of tool shock absorbers, sharp chain for chain saws, etc.); and when tools have finally worn out beyond repair they should be either traded for new tools or disposed of properly.

7.3 WORKERS COMPENSATION AND DISABILITY DETERMINATION

At this writing, some Western countries do compensate for vibration syndrome. Most recently the U.K. has recognized prescription; certain provinces of Canada compensate the worker; and Japan has begun compensation to name a few. The U.S. currently does not compensate the worker for vibration syndrome, however, private insurance carriers may at their discretion do so. The major difficulty is determining the extent of the worker's disability in the absence of reliable objective tests, thus usually the patient's medical and work histories are used of necessity. In some countries the Taylor-Pelmear classification system is used. In other countries, for example Sweden, the tests developed by Nielsen (14) are used (see Ch. 2) together with a medical-legal model developed by Gemne et al (15). This medical-legal model attempts to determine the amount of circulatory impairment, nerve impairment, and musculoskeletal disturbances the worker is afflicted with, before recommending job changes, compensation, etc.

Because vibration syndrome appears to be incurable, medical techniques are pallative at best, and with the lack of objective testing, occupational physicians struggle with the problem of when to remove the worker from the job and how to treat the malady. As an example, a prominent occupational hand-arm surgeon, Dr. Hester Hursh, who is responsible for several hundred workers in several foundries, and who has longitudinally followed these workers for several years proposes and uses the treatment format shown in Fig. 7.5 (presented at a recent vibration conference (9)). This scheme attempts to recognize and combine the time course of vibration syndrome, with the symptomatology and treatment in proportion to the extent of the malady; and most importantly to determine if and when the worker must be completely removed from exposure. Other physicians understandably use their own course of treatment and determine when the worker must leave the vibration exposed job. Thus, it should be obvious to the reader by this time that the lack of reliable objective tests and the lack of a cure for the malady makes things very difficult for the physician, the nurse, the worker, management and labor. Thus, the quest for objective tests and cures as well as an understanding of the etiology of the malady continues.

In summary, it would appear that a multifaceted approach to minimizing and/or eliminating VWF from the workplace is absolutely necessary. This should include medical evaluation, standards, A/V tools and gloves, realistic work practices, redesign of the work process if possible, and a sense of awareness by all concerned as to the hazards of VWF.

Fig. 7.5 Vibration syndrome occupational therapy regimen proposed and used by hand-arm surgeon Dr. H. Hursh in the U.S. (Courtesy of Dr. Hursh)

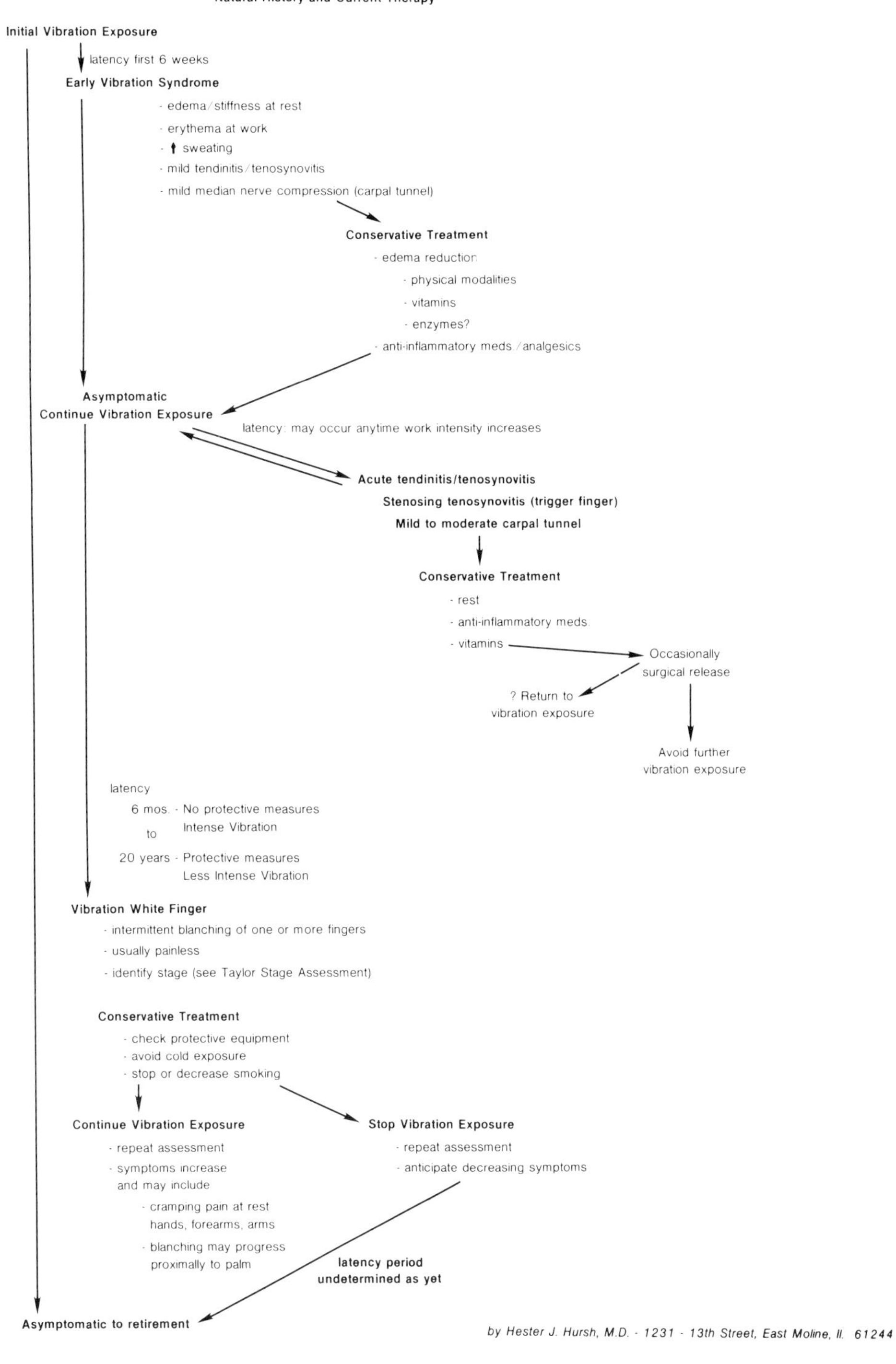

by Hester J. Hursh, M.D. - 1231 - 13th Street, East Moline, Il. 61244

7.4 WHOLE-BODY VIBRATION IN THE WORKPLACE

As the reader has surmised by now, whole-body vibration in the workplace, because of the paucity of hard epidemiological and medical data is much more elusive and not well defined as compared to hand-arm vibration. However, certain precautions can be taken to minimize worker exposure.

7.4.1 Standards and work practices guide

As previously discussed ISO 2631 with its difficulties still represents one of the very few, if not the only, such comprehensive document of its kind. However, the reader is advised to use it cautiously knowing its limitations. In addition to ISO 2631, the following work practices guide (8) should be helpful, especially if the vibrating source eminates human resonance components in the 4-8 Hz, a_z range or 1-2 Hz, a_x, a_y range: a) Limit the time spent by workers on a vibrating surface to no more than is absolutely necessary to perform the job safely; b) Have machine controls moved, wherever possible, off the vibrating surface; c) Mechanically isolate the vibrating source or the surface where the workers are stationed in order to reduce exposure; d) Compare measured vibration acceleration to ISO 2631; e) Carefully maintain vibrating machinery to prevent the development of excess vibration.

In the case of very low frequency vibration (less than 1 Hz) which brings about motion sickness, Bittner and Guignard (16) at the U.S. Naval Biodynamics Laboratory recommend the following human factors engineering approaches while investigating various workstations aboard various U.S. Coast Guard ships: a) Locate critical work stations near the ship's effective center of rotation; b) Minimize head movements; c) Align operator stations with a principal axis of the ship's hull; d) Avoid combining provocative sources; (e) Provide an external visual frame of reference at stations where seasickness may seriously impair mission effectiveness (16).

They further recommend that some of these human factors principles can be applied while a ship is being designed thereby avoid problems later after it has become operational.

7.4.2 Vibration isolation

In chapter 3, vibration theory and mathematics were presented. The question arises as to how to isolate the worker from the whole-body vibration source(s), which could be a vehicle engine, a vibrating process in a factory (e.g., mold shakeout process in a foundry), rough terrain being traversed by a heavy equipment vehicle, a farm tractor moving perpendicular to the planting rows, etc. The issue really comes down to either isolating the source, isolating the worker from the source, or both together. There have, for example, been excellent results (in some cases) of mechanically isolating a vehicle cab from the rest of the vibrating vehicle (e.g, Caterpillar Tractor Company's - cushion hitch isolater between the cab and the bowl of their heavy equipment scraper units). On the other hand, there have been examples of failures where heavy equipment vehicle cabs were vibration isolated from the engine and wheels, leaving the operator isolated and actually unable to control the vehicle because he was unable to feel the ground motion below him. In the case of the development of A/V gloves, one of the problems was (and to some extent still is), that in order to get sufficient vibration isolation for the hands and fingers, large amounts of conventional absorbing materials were needed, making the glove large and creating the situation when the worker could not grasp the tool and pressed hard so he could get a grip on the tool, thereby increasing the vibration coupling between the tool and his hand and defeating the purpose of the glove. Newer, thinner A/V glove materials helped reduce this problem. There are many companies, such as the Bostrom Seat Co. of Milwaukee, Wisconsin who have given thousands of engineering hours to developing antivibration passive seats for vehicle operators (see Fig. 7.6). These seats in many cases require the operator to adjust the seat suspension system to his weight and on the main they work relatively well except at the very low frequencies when the suspension system can only partially attenuate these frequencies.

In an effort to overcome some of the problems of passive A/V seats the early 1970's saw the design of "active seats" using a servocontrol system. The seat was instrumented with either an accelerometer, velocity or displacement transducer, electronically conditioned, and subsequently fed to a hydraulic actuator at the base of the seat. In theory what was to happen was that incoming vibration was sensed, inverted and electronically summed, and "fed back" to the seat surface in an effort to quickly "null out" the impinging vibration. In practice what happened was: a) The seats were very costly; b) They were unreliable and actually (in some cases) exacerbated the problem, because the seats themselves became hydraulic shakers and thus the operators were exposed to more vibration than originally impinging on them! Thus these active vibration seats were removed from the market and passive seat designs remain. The important thing to remember is that there are no magic engineering solutions, some work, some do not work, some work better than others. What is presented next is a chronological listing of approaches and steps vibration engineers tend to use to reach solutions in typical vibration problems. This will aid in the reader's understanding of how solutions are arrived at with specifics available in advanced texts on the subject (17-20).

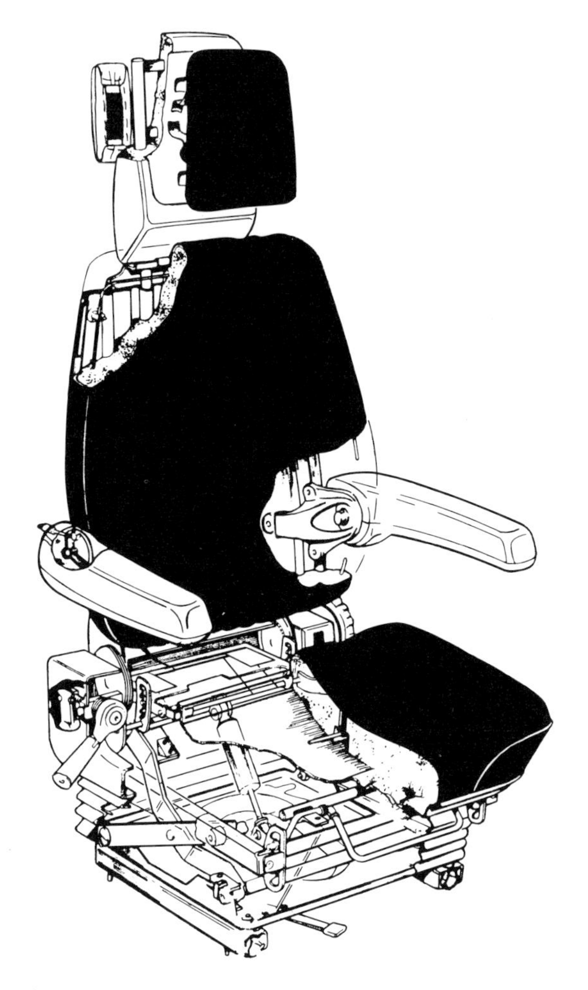

Fig. 7.6 Cutaway view of a typical vehicle passive suspension seat. (Courtesy Bostrom Co.)

First, the vibrating machine, system, etc. is visually examined. Second, accelerometers and other transducers are attached to various critical points on the vibrating system, data is obtained, and FM tape recorded. Third, in some cases where a system reacts to vibration rather than a source itself, it is necessary to "excite" the system either by exciting it with a sweeping vibration shaker or tapping various places on the system with a specially instrumented hammer. In both cases accelerometers are placed at critical points to detect the systems response (and resonances) to the vibration excitation, followed by FM tape recording of these data. Fourth, the system is mathematically modeled using appropriate equations of motion. Fifth, the aforementioned tape recorded data is digitized and computer analyzed for vibration spectrum, correlation, resonances, etc. with the attempt being to use these data to supply critical information (e.g., damping, spring constants, etc.) in solving these modeling equations. Sixth, once critical parameters of the equations of motion are solved, the engineer is next faced with the problem of appropriately removing the unwanted vibration from the system; he can use appropriate materials to filter out unwanted vibration at specific points on the system; he can vibration isolate the system from a floor for example by using appropriate shock and vibration isolators or actually mounting the unit on a mechanically floating concrete pad in a pit in the ground, for example; he can place appropriate mechanical filters between vibrating parts, etc. It all depends on the problem; he can, for example, shift the systems resonances higher in frequency by stiffening up a system mechanically and then once this achieved, he can filter out these higher frequencies then he could at lower frequencies. Seventh, he modifies the system on the computer, later followed by actual modifications on site. Finally, he must next determine if his modifications solved the problem under actual operating conditions so steps two through five are repeated and the data analyzed. If an appropriate solution has been found, then it ends there; if not the engineering continues.

In summary each vibration problem is unique and limited usually to the tool or machine being analyzed under a particular working condition. Usually, vibration engineers obtain and record measurements of a vibrating source at various locations on the source including points where the human comes into contact with the source. Fourier analysis, modal analysis, transfer function analysis, and mechanical impedance analysis (to name a few) are used on the data in an effort to "fingerprint" effectively the vibrating source under the given work conditions. Equations of motion are developed for the given condition, noting system resonance points. Finally, with a knowledge of damping materials and damping techniques, these are applied to the system under test, comparing the theoretical reductions to actual reductions obtained. Thus vibration reduction becomes a repetitive and difficult process until the desired solution is obtained. If damping materials and techniques are insufficient, then various mechanical changes are usually made to the vibrating system (e.g., stiffening the mechanical system, reducing the system mass) in an effort to shift resonance points higher in frequency where damping materials become more effective. (21)

REFERENCES

1. D.E. Wasserman, Raynaud's Phenomenon as It Relates to Hand-Tool Vibration in the Workplace, J. Amer. Ind. Hyg. Assoc. 46 (1985) 10-18.
2. S. Fishman, D.E. Wasserman, and V. Behrens, Vibration Syndrome (CIB #38), DHHS/NIOSH Public. No. 83-110, 1983.
3. W. Taylor and D.E. Wasserman, Background on Vibration, ACGIH Documentation of Threshold Limit Values for Physical Agents in the Work Environment, 1981.
4. D.E. Wasserman and W. Taylor (Eds.), Proceedings of the International Occupational Hand-Arm Vibration Conference, DHEW/NIOSH Public. No. 77-170, 1977.
5. A.J. Brammer and W. Taylor (Eds.), Vibration Effects on the Hand and Arm in Industry, John Wiley and Sons Publishers, New York, 1982.
6. A. Hamilton, A Study of Spastic Anemia in the Hands of Stonecutters: An Effect of the Air Hammer on the Hands of Stonecutters, Industrial Accidents and Hygiene Series, Bulletin 236, No. 19, U.S.D.O.L., Bureau of Labor Statistics, 1918.
7. W. Taylor, D. Wasserman, V. Behrens, D. Reynolds and S. Samueloff, Effect of the Air Hammer on the Hands of Stonecutters. The Limestone Quarries of Bedford, Indiana, Revisited, Brit. J. of Ind. Medicine, 41 (1984) 289-295.
8. Proceedings of the Symposium on Occupational Health Hazard Control Technology in the Foundry and Secondary Non-Ferrous Smelting Industries, DHHS/NIOSH Public. No. 81-114, 1981.
9. Proceedings of the American Foundrymen's Society/American Occupational Medical Association Vibration White Finger Conference, Cinti., Ohio, June 4-5, 1985.
10. V.K. Goel and K. Rim, Role of Gloves in Reducing Vibration: An Analysis for Pneumatic Chipping Hammer, J. Amer. Hyg. Assoc. (In Press).
11. D. Wasserman, W. Taylor, V. Behrens, S. Samueloff and D. Reynolds, VWF Disease in U.S. Workers Using Pneumatic Chipping and Grinding Hand Tools - Vol. I Epidemiology, DHHS/NIOSH Public. No. 82-118, 1982.
12. Vibration Syndrome, NIOSH Video Tape 177, TV Dept. DHHS/PMS/CDC/NIOSH, 4676 Columbia Parkway, Cincinnati, Ohio 45226 (27 minutes, produced in 1981).
13. V. Behrens, D. Wasserman, W. Taylor and T. Wilcox, Vibration Syndrome in Chipping and Grinding Workers, J. Occup. Medicine (Special Supplement), 26 (1984) 765-788.
14. S.L. Nielsen, Raynaud's Phenomena and Finger Systolic Pressure During Cooling, Scand. J. Clin. Labl Invest., 38 (1978) 765-770.
15. G. Gemne, L. Ekenvail, J.E. Hansson and I.M. Lidstrom, A Medical Legal Model for the Evaluation of Harmful Influences from Hand-Arm Vibration, Arbeit. Halsa, 2 (1986) 1-26.
16. A.C. Bittner and J.C. Guignard, Human Factors Engineering Principles for Minimizing Adverse Ship Motion Effects: Theory and Practice, Naval Engrs. Journal, 97 (1985) 205-213.
17. F.S. Tse, I.E. Morse and R.T. Hinkle, Mechanical Vibrations, Allyn & Bacon Publ., Boston, 1971.
18. A.D. Nashif, D.I.G. Jones and J.P. Henderson, Vibration Damping, John Wiley and Sons Publ., New York, 1985.
19. D.J. Ewing, Modal Testing: Theory and Practice, John Wiley and Sons Publ., New York, 1985.
20. R.A. Ibrahim, Parametric Random Vibration, John Wiley and Sons Publ., New York, 1985.
21. D.E. Wasserman, Motion and Vibration - Chapter 6, Handbook of Human Factors/Ergonomics, John Wiley and Sons Publ., New York, 1986.

SUPPLEMENTARY BIBLIOGRAPHY

L. Greenberg and D.B. Chaffin, Workers and Their Tools, A Guide to the Ergonomic Design of Hand-Tools and Small Presses, Univ. of Michigan Dept. of Engr. Report, 1975.

A Henschel, Recommendations for Control of Occupational Safety and Health Hazards in Foundries, DHHS/NIOSH Public. No. 85-116, 1985.

J.E. Hansson and S. Kihlberg, A Test Rig for the Measurement of Vibration in Hand-Held Power Tools, Applied Ergonomics, 14 (1983) 11-18.

E. Christ, Reduced hand-arm Vibration Load Due to Protective Gloves, Trade Union Institute for Occupational Safety Report, St. Augustia, German, 1982.

L. Gidlund and B. Linguist, Age-old Method Gets a New Lease on Life, Atlas-Copco Co. Technical Report, Stockholm, 1982.

N. Olsen and S.L.Nielsen, Diagnosis of Raynaud's Phenomenon in Quarrymen's Traumatic Vasospastic Disease, Scand. J. Work-Envir. - Health, 5 (1979) 249-256.

S.W. Vines, Vibration White Finger - Implications for Occupational Health Nurses, Occupational Health Nursing 5 (1984) 526-529.

F.S. Keefe, R.S. Surwitt, and R.N. Pilon, Biofeedback Antogenic Training, and Progressive Relaxation in the Treatment of Raynaud's Disease: A Comparative Study, J. Applied Behavioral Analysis, 13 (1980) 3-11.

M.J. Griffin, C.R. Macfarlane and C.D. Norman, The Transmission of Vibration to the Hand and the Influence of Gloves in Vibration Effects on the Hand and Arm in Industry (A.J. Brammer and W. Taylor, Eds.), John Wiley and Sons Publ., New York, 1982.

Chapter 8

Another View of Vibration

8.1 INTRODUCTION

In this the final chapter, we ask an usual question: Is all mechanical vibration harmful or can it be harnessed and used in a positive way? This same question has been asked with regard to ionizing radiation, etc. In this author's view, under certain and controlled circumstances mechanical vibration can be used in a positive way. Rephrased, can vibration be used perhaps therapeutically to reduce disease or in other useful ways? Unfortunately, the literature is very sparse in this area of vibration.

8.2 VIBRATION AS A POSSIBLE THERAPEUTIC MODALITY IN REDUCING DISUSE OSTEOPOROSIS

In the U.S. there are some 500,000 spinal cord injured (SCI) patients with an incidence rate of some 10-15,000 yearly. These SCI victims became paralyzed due to vehicle accidents, sports accidents, shootings, stab wounds, etc. These patients are either quadriplegics and paralyzed in all four limbs, or paraplegics who are paralyzed only in the lower limbs. Collectively they are called "paralytics." When a traumatic injury occurs and a paralysis results not only is the patient paralyzed, but several other insidious problems occur: a) the muscles start to atrophy in the unused limbs; b) the cardiovascular and pulmonary systems become deconditioned; c) the patient cannot sweat below the level of injury thus increasing their chances of heat stroke or exhaustion on hot days as during exercise; d) disuse osteoporosis also called "bone thinning or bone dissolution" sets in the paralyzed limbs resulting in large amounts of calcium being excreted by the kidneys. Disuse osteoporosis is not simply solved, feeding the patients calcium is dangerous (because kidney stones can occur) and in of itself may do little because the body tends to excrete this calcium, too (1,2). This same situation of disuse osteoporosis plagues astronauts returning to earth from a zero gravity situation and affects thousands of older females, too.

The use of active physical therapy using computer controlled, closed loop functional electrical stimulation has helped in the cessation of and the rebuilding of muscle, in ambulation, and reversing cardiomuscular pulmonary deconditioning, but it has been less successful in stopping disuse osteoporosis (1-3).

8.2.1 Osteoporosis in paralytics

According to Wolff's Law of Bone Transformation (4,5) the deposition of hard bone is related to the application of stresses to the bones, of an amplitude and in the geometry of stresses applied in standing and walking in normal persons. The opposite phenomenon (bone dissolution) depends on the availability to the bone of the materials (calcium, phosphate, and carbonate ions) involved in the metabolism of bone. In the living body, the condition of the bone depends on an equilibrium between the rate of deposition and the rate of dissolution.

In the paralyzed person, there are few if any force stresses applied to the long bones of the lower limbs, so that reduced bone deposition occurs. The result of this is that bone undergoes a net dissolution of osteoporosis. This osteoporosis becomes severe enough that the affected bones are weakened and fractures are common in paralytics and as many as 10% of paralyzed persons experience repeated breakages (6).

Therefore, there is an ever present hazard of bone fracture which must be considered in first handling paralyzed humans. Benassy (7) observed that bone healing was not a problem in paralytics, which implies that their primary difficulty is associated with the osteoporosis per se and the lack of bone strength associated with it.

Guttman (8) and Sunderland (9) pointed out the importance of osteoporosis associated with the forced inactivity to the welfare of paralyzed patients.

The mechanisms of the effect of stresses on bone deposition are not known at this time. One suggested mechanism is that bone is deposited along electric current or voltage lines, and that such electrical fields may be generated by piezoelectric effect when stresses are applied to the bone (10,11).

Another suggestion for combatting osteoporosis is the application of stresses by subjecting the limbs to artificial g forces using controlled continues mechanical vibration in one or more axis. Such studies would need to: a) test the hypothesis that vibration does reverse osteoporosis; b) determine the optimal amounts and frequencies of vibration to be applied; and c) determine the optimal duration and frequency of repetition of exposure to the vibration.

Recent NASA studies in astronauts subjected to zero gravity have indicated that detectable osteoporosis occurs in their leg bones with remarkably brief exposures to low gravity. These studies would suggest that prevention of osteoporosis and/or reversal of the osteoporosis may require the application of forces to the bone that approximate those encountered in normal walking.

In a recent study at the NCRE laboratory by Phillips, Petrofsky, et al (3), two paraplegic subjects legs were exposed to 3 g_{peak} (vertical) sinusoidal acceleration (corresponding to 2.1 g_{rms} acceleration) at 15 Hz for 3 alternate days/week for 12 weeks. Each subject sat in a cushioned chair, one leg was tightly coupled at 90 degrees to an electrodynamic shaker, the other acted as a control and was not on the shaker. In this case the subjects long bones (tibia and fibula) were exposed to vibration. Both subjects had at least 75% bone demineralization and showed a 53.5% increase in bone mineral of the proximal tibia, but a 7.5% bone loss in the distal femur (see Fig. 8.1). Thus it would appear in this preliminary study that mechanical vibration might have potential in this area, much work needs to be done, and it is continuing at NCRE (12). Finally on this subject, an extremely difficult problem is the quantitative measurement of the degree of osteoporosis. The technique we have available at this time involves the use of CAT scans (Computerized Axial Tomography), which is expensive and not always immediately available. A computer analysis of CAT scan results is required to make the resulting data quantitative for bone density.

A promising development is a technique for the same kind of scanning, using gamma rays from Iodine-135. Radiation exposures are much smaller than with conventional scanners, and there is indication that the accuracy of estimation of bone density will be greater (13).

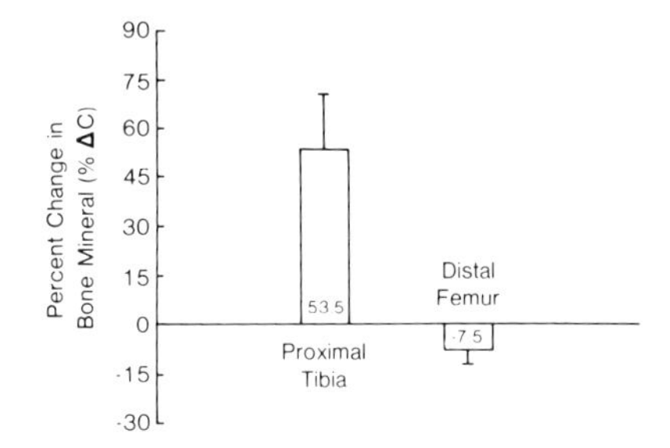

Fig. 8.1 Change in bone mineral density in the lower leg of two spinal cord injured patients after three months of vibration exposure of 15 Hz, 3 g_z peak.
(From Phillips, Petrofsky et al, reference (3))

The reader should be aware that for several years static and dynamic (e.g., vibration) loading of bone has been ongoing by biomechanical engineers and scientists to study such parameters as bone strength, stress and strain, shear forces, resonances, etc. using cadavers, animals, etc. (14-18). However, the therapeutic use of vibration on living humans as described above (in vivo) is relatively new and unique.

8.3 VIBRATION AS A POSSIBLE THERAPEUTIC MODALITY IN PAIN MANAGEMENT

There are many ways to manage pain, e.g., the use of drugs, soothing baths, ice, TENS (Transcutaneous Electrical Nerve Stimulation), wet heat, etc. A few researchers have attempted to use sinusoidal vibration in the 50-200 Hz range at very low amplitudes (19,20). Some of this work appears to be an extension of previous studies on tactile stimulation and perception. Vibration is applied locally to specific pain sites. In one of these studies (19), 366 pain patients were treated with vibration, 256 obtained pain relief in sessions from 25-45 minutes in length, 2-3 sessions/week. Pain relief ranged from 3-12 hours in duration depending on when the pain probe was applied. The vibration frequencies varied from 5-200 Hz, depending on the probe pressure (i.e., coupling) between it and the body. Frequencies below 50 Hz had no significant effect on pain reduction (19-20).

8.4 VIBRATION AS A POSSIBLE THERAPEUTIC MODALITY IN REDUCING CARDIOVASCULAR DECONDITIONING AND DISFUNCTION

For approximately 10 years U.S. researchers Bhattacharya, Knapp, McCutcheon and associates have attempted to use whole-body vibration as a cardiac assist modality by synchronizing externally applied vibration to events in the normal cardiac cycle (22-24). Similarily, they have tried to reduce cardiovascular deconditioning (experienced by astronauts) by using whole-body vibration as an analog for weightlessness (25). Among other things, these researchers have found that cardiomuscular and musculoskeletal conditioning can be achieved by using an oscillating bed-like shaker to simulate G profiles of normal adults. They have also developed the basic mathematical solutions between externally applied whole-body vibration and the (animal) subjects as well as the effects of phase synchronization (and lack of synchronization) between the externally applied vibration and the subjects cardiac cycle.

Admittedly this work is esoteric, but nonetheless dramatic on the positive potential of vibration as a therapeutic modality.

8.5 A FEW OTHER POTENTIAL POSITIVE USES OF VIBRATION

In an effort to test new dental materials, multiaxis servohydraulic shakers have been successfully used to simulate jaw motion and wear on new dental materials (26,27). Doing so reduces the number of actual clinical trials needed to perfect the materials.

Finally, as an example of a safety device, a Japanese invention (28) uses a small vibrator in the seat back of the driver's seat. If the parking brake has not been properly released and the driver attempts to drive away with the brake engaged, the electromagnetic vibrator is actuated warning the driver of the problem. It deactivates after the brake has been released.

In summary, in this final chapter we have attempted to briefly demonstrate that there are potential positive human uses of controlled vibration, unfortunately, the workplace situation contains little or no such vibration and thus we must be constantly aware of the negative and potentially negative effects of uncontrolled vibration in the workplace.

We conclude this book on this note; the original premise for writing this book was to inform you the reader of the varied and complex aspects of occupational vibration, and to act as a catalyst for you to pursue your individual problem area armed with a broad sense of understanding of vibration. Those of us working in this area are very lucky because there is in a single depository the largest vibration literature collection in the world, where virtually everyone publishing papers, books, etc. sends a copy of their work. This collection is maintained and constantly updated under the supervision of an internationally known colleague, Dr. Michael Griffin. Should you have very specific needs for published materials in virtually any aspect of vibration, you may wish to directly communicate with him at the following address:

Dr. Michael Griffin
Institute of Sound & Vibration Research
University of Southampton
Southampton, United Kingdom S09-5NH

REFERENCES

1. J.S. Petrofsky and C.A. Phillips, Active Physical Therapy - A Modern Approach to Rehabilitation Therapy, J. Neuro. Orthop. Surg. 4 (1983) 165-173.
2. J.S. Petrofsky, C.A. Phillips and M. Sawka, Muscle Fiber Recruitment and Blood Pressure Response to Isometric Exercise, J. Appl. Physiol. 50 (1980) 32-37.
3. C.A. Phillips, J.S. Petrofsky, D.M. Hendershot and D. Stafford, Functional Electrical Exercise - A Comprehensive Approach for Physical Conditioning of the Spinal Cord Injured Patient, Orthopedics 7 (1984) 1112-1123.
4. R.W. Stacy, D. Williams, R. Worden and W. McMorris, Essentials of Biological and Medical Physics, McGraw-Hill Book Co., New York, 1955.
5. R.W. Treharne, Review of Wolff's Law and Its Proposed Means of Operation, Orthopaedic Review 10 (1981) 35-47.
6. K.T. Ragnarsson and G.H. Sell, Lower Extremity Fractures After Spinal Cord Injury: A Retrospective Study, Arch. Phys. Med. Rehab. 62 (1981) 418-423.
7. J. Bennassy, Associated Fractures of the Limbs in Traumatic Paraplegia and Tetraplegia, Paraplegia 5 (1968) 209-211.
8. L. Guttman, Spinal Cord Injuries: Comprehensive Managements and Research, (2nd. Ed.) Blackwell Scientific Publishers, Oxford, England, 1967.

9. S. Sunderland, Nerves and Nerve Injuries, Williams and Wilkins Publ. Co., Baltimore, 1968.
10. C. Bassett and R.O. Becker, Generation of Electric Potentials by Bone in Response to Mechanical Stress, Science 137 (1962) 1063-1064.
11. C. Bassett and M.G. Valde, Modification of Fracture Repair with Selected Pulsing Magnetic Fields, J. Bone Joint Surgery 64 (1982) 888-895.
12. D.E. Wasserman, C. Phillips, J.S. Petrofsky, The Potential Therapeutic Effects of Segmental Vibration on Osteoporosis, Proceedings of the Twelfth Congress on Acoustics, Toronto, Canada, July 1986.
13. T.N. Hangartner and T.R. Overton, Quantitative Measurement of Bone Density Using Gamma-Ray Computed Tomography, J. Computer Assisted Tomography, 6 (1982) 1156-1162.
14. J.M. Jurist, In Vivo Determination of the Elastic Response of Bone, Phys, Med. Biol. 15 (1970) 427-434.
15. W.P. Doherty, E.G. Bovill and E.L. Wilson, Evaluation of the Use of Resonant Frequencies to Characterize Physical Properties of Human Long Bones, J. Biomechanics 7 (1974) 559-561.
16. L.E. Lauzon and C.T. Rubin, Static vs Dynamic Loads as an Influence on Bone Remodelling, J. Biomechanics 17 (1980) 897-905.
17. J.F. Laferty and P.V.V. Raju, The Influence of Stress Frequency on the Fatigue Strength of Cortical Bone, Trans. of the Amer. Soc. Mech. Engr. 101 (1979) 112-113.
18. P.V. Spiegl and J.M. Jurist, Prediction of Ulnar Resonant Frequency, J. Biomechanics 8 (1975) 213-217.
19. T. Lunderberg, R. Nordemar and D. Ottoson, Pain Alleviation by Vibratory Stimulation, Pain 20 (1984) 25-44.
20. R. Melzack and B. Schecter, Itch and Vibration, Science 147 (1965) 1047-1098.
21. P.D. Wall and J.R. Cronley-Dillon, Pain, Itch and Vibration, Arch. Neurol. 2 (1960) 365-375.
22. A. Bhattacharya, C.F. Knapp, D.P. McCutcheon and J.M. Evans, Modification of Cardiac Function by Synchronized Oscillating Acceleration, J. Appl. Physiol. 47 (1979) 612-620.
23. A. Bhattacharya, C.F. Knapp, E.P. McCutcheon and J.M. Evans, Cardiac Responses of Dogs to Nonsynchronomal and Heart Synchronous Whole-Body Vibration, J. Appl. Physiol. 46 (1979) 549-555.
24. A. Bhattacharya, C.F. Knapp, E.P. McCutcheon and R.G. Edwards, Parameters for Assessing Vibration-Induced Cardiovascular Responses in Awake Dogs, J. Appl. Physiol. 42 (1977) 682-689.
25. A. Bhattacharya, C.F. Knapp, E.P. McCutcheon, J. Kearney and A. Cornish, Effects of Whole-Body Oscillating Acceleration on Orthostatic Response After Dynamically Imposed Oscillating Forces for Preventing Cardiovascular Deconditioning, Proceedings of the First International Conference on Mechanisms in Medicine and Biology, 1977.
26. R. Delong and W.H. Douglas, Development of an Artificial Oral Environment for the Testing of Dental Restoratives: Bi-axial Force and Movement Control, J. Dental Research 62 (1983) 32-36.
27. C.M. Gibbs, T. Messerman and J.B. Reswick, The Case Ghattic Replication for the Investigation of Mandibular Movements, Case Western Reserve Univ. Engr. Report EDC 4-66-14, Cleveland, Ohio 1966.
28. S. Yamazaki, Safety Device for Vehicle Seat with a Vibrator, U.S. Patient No. 9,465,158 (Toyota Motor Co.), Aug. 14, 1984

Appendix

(Selected exerpts from: Industrial Accidents and Hygiene Series, Bulletin 236, Number 19.) Effect of the Air Hammer on the Hands of Stonecutters

by

Alice Hamilton, M.D.

United States Bureau of Labor Statistics, July 1918

REPORTS OF PHYSICIANS FOR THE BUREAU OF LABOR STATISTICS.

A STUDY OF SPASTIC ANEMIA IN THE HANDS OF STONECUTTERS

By Alice Hamilton, M.D.

In March, 1917, Dr. Joseph L. Miller, of Chicago, called my attention to an unusual occupational disease in a stonecutter from the Indiana limestone region, who was accustomed to use the air hammer in his work. The man complained of attacks of numbness and blanching in certain finger, coming on suddenly under the influence of cold and then disappearing again. When he called on Dr. Miller for advice he had no visible signs of the trouble, but from the description he gave Dr. Miller believed it to be a form of Raynaud's disease and he thought that since there were said to be many similar cases in the limestone region it was a condition that merited close study. The Bureau of Labor Statistics had made a preliminary inquiry into the matter and found that the limestone cutters do suffer from "dead fingers." Later is was reported that granite and marble cutters do not suffer from this affection. This was denied by the men. The bureau authorized me to visit the limestone belt of Indiana, the granite-cutting centers in Quincy, Mass., and Barr, Vt., the Marble shops of Proctor, Vt., Long Island City, and Baltimore, and the sandstone mills of northern Ohio.

Since the condition in the men's hands which was the object of this inquiry comes on under the influence of cold, I made my visits during January, February, and March of 1918 on the days when the temperature was between 14 degrees F. and and 34 degrees F. I discovered a very clearly defined localized anemia of certain fingers which is undoubtedly associated with the use of the air hammer and which, while it lasts, makes the fingers numb and clumsy, causing the man more or less discomfort and sometimes hampering his work.

The pneumatic hammer consists of a handle containing the hammer, which is driven by compressed air and is said to deliver from 3,000 to 3,500 strokes a minute. The amount of air delivered through a hammer can be controlled by a valve in the pipe conveying the air, and the air escapes through an exhaust opening in the handle itself. This handle is held in the right hand in various ways, sometimes with the palm of the hand down and all the fingers grasping the handle equally, sometimes with the palm up and the grasp exerted chiefly by the thumb, index, and middle fingers: it may also be held between the thumb, index, and middle fingers, very much as a pen is held. Hammers are of various sizes; there is the small half-inch, the medium five-eighths or three-fourths inch, and the large 1-inch hammer. The tool (the chisel) is held by the left hand against the hammer, with the cutting edge pressed against the stone. Italian workmen usually slip the tool between the little and ring fingers, so that it rests against the side of the little finger, where a large callus develops. Other workmen grasp the tool with all four fingers. In either case the little and ring fingers, being nearest the cutting end of the tool, are pressed most closely against it in order to guide it.

The conditions under which stone is cut differ somewhat for the four kinds of stone. In the limestone region of Indiana and in the sandstone region of Ohio there are large mills heated somewhat in winter, so that the temperature is perhaps 10 or, rarely, 20 degrees higher indoors than outside. This would mean that when the thermometer stands at 15 degrees F. the working atmosphere will be at about freezing point or perhaps as high as 38 degrees F. In Quincy and in Barre, granite is cut in sheds, which in the former town are wide open, but in Barre are inclosed and sometimes slightly warmed. They are, however, colder than the western mills and in very cold weather work has to be suspended. The marble shops are inclosed and usually better heated than any of the other stonecutting shops I visited. I did not see a place in any mill or shed where a man could warm his hands conveniently.

The air hammer is used in cutting all four kinds of stone, but not to the same extent in all. Limestone cutters use it almost all the time. When one enters a mill in the limestone region the stonecutters, with a very few exceptions, are all seen to be using the air hammer. It is rare to see more than two or three men wielding the mallet, unless they are apprentices who are required to use it.

In cutting limestone the air hammer can be used both for shaping the block of 'stone, a process known as roughing out, and for cleaning up or making a smooth surface. Many men say that the roughing out really be done with the mallet, but in practice the air hammer is used. Limestone cutters use all sizes of tools; the carvers use the smaller ones chiefly or entirely. Marble cutters come next in their use of the air hammer. They work more with the mallet than do the limestone men, but the greater part of their work is the pneumatic tool and usually the smaller sizes, the half or three-fourths inch tool.

Granite cutters can not use this tool for shaping the block. That must be done by hand, because the stone is so hard. For dressing the surface they use two machines, a large heavy surfacer with a big handle, which is grasped in both hands and held upright, the tool pressing on the surface of the stone.

The tool in both these machines is held in place by the hammer and stone never grasped or guided by the left hand. For lettering and carving, however, the granite worker uses the same sort of air hammer as is found in marble and limestone mills, and there are granite workers who use it all day long, but these are the exception. As a rule the men I questioned in the granite sheds use it only four, five, or six hours a day.

In sandstone the air hammer seems to be of little use. A mill I visited near Amherst, Ohio, had five air hammers for 30 men, and that number was quite sufficient. Sandstone does not require much tooling. It is used chiefly for paving stone, curbstones, grindstones, and exterior building stone. Much of the tooling required is done by hand, for the nature of the stone makes work with the air hammer difficult or impossible. I questioned 15 sandstone cutters and was told by 6 that they had never used the air hammer at all. Two had formerly used it in marblework, but not in sandstone, and 7 used it now and then for sandstone, but hardly more than half an hour during the day.

DESCRIPTION OF STONECUTTERS' WHITE FINGERS

The trouble which I found among the limestone cutters, granite workers, and marble carvers is not Raynaud's disease. It is a spastic anemia, affecting the arterioles of the fingers and hands--it comes on in sudden and recurring attacks under the stimulus of cold as does Raynaud's syncope, and the pallor is much more pronounced, but it is not succeeded by the stage of extreme asphyxia so characteristic of Raynaud's--it is not symmetrical, even when in both hands, and it does not result in the wasting and death of tissue which accompany Raynaud's. I did not see any marked swelling or excessive congestion, or severe pain. Nor did I see hands with thickened fingers or deformed nails; on the contrary these stonecutters had well-formed and well-preserved nails such as one seldom sees in manual workers. No history was given me of necrosis or blisters or ulcers or desquamating skin, and the pain complained of was nothing like so severe as that often experienced in Raynaud's disease. As to the distribution of the vascular spasm, this is not capricious as in Raynaud's but apparently determined by definite causes. It is not really symmetrical. The left hand is usually the only one affected, and when the right hand suffers the fingers involved are not the ones corresponding to those involved on the left hand.

The examination which I was able to give the men in the absence of all laboratory equipment established the fact of a spastic anemia undoubtedly of occupational origin and accompanied by certain more or less definite sensory changes which sometimes persist to a lesser extent between the attacks. I could not determine how great the latter are nor whether there is any real impairment of the function of the affected hand. That part of the investigation was undertaken in Chicago by Dr. Thor Rothstein.

A description of one or two of the more marked cases will show just what this condition is. The first is a limestone cutter whom I saw early in the morning when the temperature was about 14 degrees F. He had been out of doors for over half an hour and in order to be able to show me his hands in a typical condition he had refrained from rubbing them violently and swinging his arms about, as he would ordinarily do to restore the circulation. The discomfort, however, had grown so intense in his fingers that he could not bear it any longer and almost at once after I arrived he began rubbing and kneading and shaking his hands. The four fingers of his left hand were a dead, greenish white and were shrunken, quite like the hand of a corpse. The whiteness involved all the little finger to the knuckle, but in the other fingers it stopped midway between knuckle and second joint. As he rubbed his hand the contrast between fingers and hand increased and at one stage it was very striking, the crimson and slight swollen hand meeting the white shrunken fingers abruptly, without any intermediate zone. On the palmar side the condition was not so distinct, for the skin was too thick and calloused to allow the color to show well.

The right hand was much less affected, the little finger escaped altogether, the three others were white, but not dead white, as far as the second joints, and there was a ring of white around the second phalanx of the thumb. After vigorous massage and beating of his arms back and forth over his chest, the blood gradually filled the fingers and the appearance then was fairly normal, showing only a moderately purplish red color, and no swelling.

This man is 39 years old and has cut stone for 22 years. While using the ordinary tools of his trade he had no trouble of this kind. Nine years ago he began to work with the air hammer and during the second winter after that he noticed that the ring finger of the left hand had begun to "go white." Gradually the little finger became involved, then the others, and, to a less extent, the fingers of the right hand. The trouble has progressed through the years and is still increasing. There is a good deal of pain in the fingers, especially on a cold morning. As long as the dead-white condition lasts there is no real pain, but discomfort enough to make him stop work and get the blood back into his fingers for the stroke of the hammer on the tool he holds in the left hand is peculiarly intolerable when the fingers are white. As the blood comes back there is some sharp pain but it does not last. At no time, however, does the left hand feel quite natural, he is always conscious of it; indeed his whole left side, including the foot,feels differently from the right. If he holds his hands up for a few minutes they grow numb and this is annoying when he tries to read a newspaper and must continually put it down to coax the blood back into his hands. He has lost sensitiveness in the fingers, so that he can not put his left hand in his pocket and pick out a coin by the touch; he must look at it to see if it is a dime or a nickel. He is clumsy in the morning when buttoning his clothes and locking his boots. If he works all day with he hammer he has a restless, disturbed night.

The second man is a marble cutter who has followed his trade for 20 years and has had trouble with this fingers from the fifth year on. He uses the small tool almost entirely. The four fingers of the left hand were white, the little finger over the whole extent, the next two over the two distal joints, the index over the first joint. On the right hand the tips of all four fingers were white and there were irregular streaks of white along the index and middle fingers. This man complained of the pain in his fingers both when they were white and when the blood first began to come back, but his chief complaint was of nervousness from the vibration of the hammer; he said it upset him, made him "as nervous as a kitten" spoiled his sleep, made him irritable. Though he is troubled chiefly in winter, he can not put his hands in cold water in summer without making his fingers "go white." In winter, if he is working indoors he is not really hampered by the numbness in his fingers, but he can not do any fine work out of doors if the weather is at all cold, for the numbness makes his fingers clumsy.

The third is a granite cutter who has used the air hammer for 18 years and who began to feel the effects in his fingers after two years. Now his left hand shows all of the little, ring, and middle fingers involved and all but one-third of the index finger. On the right hand most of the index and middle fingers and the tips of the ring and little fingers are blanched, but not so strikingly so as the fingers of the left hand. He is "bothered" a good deal by the numbness in winter, and it comes on whenever he handles a cold tool, after which he finds it hard to do any fine work until he has managed to get the circulation started again.

There is no need to multiply these descriptions. With a few variations the men from whom full histories could be obtained told much the same tale, though the majority had not suffered as much discomfort as had these men. These stonecutters are exceptionally good material for such a study, for they are intelligent men, usually of good education and able to note and describe their symptoms clearly. There is among some of them a tendency to dwell perhaps too much on the nervous disorders which they believe are caused by the tiring vibrations of the hammer, and which give them a good deal of worry. Of nearly all of the men this is not true. Many of them have no complaint at all, except of the actual condition in the hands, but others suffer from more or less distressing symptoms which they think are caused by the vibrating hammer. The most common symptom is covered by that vague term "nervousness." They say that they feel jumpy and irritable, upset by a slamming door, and unable to settle down after a full day's work with the tool. Their sleep is disturbed and restless, and they have buzzing or ringing in the ears. The numbness in the hands is inconvenient, for they can not hold a newspaper or a book for any length of time without being forced to put it down and rub and knead their hands. Sometimes they have to sleep with the left arm hanging down from the bed or the numbness will waken them, and then they must get up and swing the arms about or bathe the hands in hot water. Some of them are not troubled at all in summer, other get numb fingers on chilly days or if they put their hands in cold water. A few men complain of trouble with the left foot, which is colder than the right.

The local changes are surprisingly similar in all these cases. There is no capricious localization of the anemia. Invariably the little finger of the left hand is affected, never, so far as I say, the left thumb, though a few men said they believed it did sometimes "go dead." If one finger on the left hand escapes, it is the index finger. The most usual distribution is over the little and ring fingers, part of the middle finger, and somewhat less of the index on the left hand, so that the line between the pale and red portions runs diagonally from the knuckle of the little finger to the first joint of the index. Sometimes there is a spot of white in the center of the left palm. The right often escapes entirely, so far as evident signs go, and when affected it is less strikingly so than the left, and the anemia is less uniform in its distribution. The right thumb is sometimes involved, the index finger often, but the other fingers are hardly ever pale except at the tips.

There were two men who surprised me by displaying left hands with anemic finger tips and right hands with anemia of all four fingers, but both proved to be left-handed and they use their hands accordingly, holding the tool in the right hand and the hammer in the left. Many men told me that the white area on their hands sometimes extended as far as the wrists on the ulnar side, but I never saw it reach even quite to the knuckles.

Usually the men say that the blanching of the fingers is most marked early in the morning and passes away more or less while they are at work with the hammer, returning in the after noon when work is over. I examined a little over 100 soft-stone, marble, and granite cutters at work and could actually see the pale fingers in only 22, but among 23 men I saw away from the plants 16 showed the condition quite strikingly after they had exposed their hands to the cold for a few minutes. Among those who came to see me at the hotel there were some whose hands grew warm fairly quickly and appeared quite normal after they had been in the room a short time, but in other cases a decided difference persisted in the color and temperature of some of the fingers of the left hand as compared with other parts of that hand and with the right hand. Several men told me that there was always some difference in the feeling of the two hands. Men who had had attacks of white fingers could always bring the condition on by washing their hands in cold water or going out of doors for a while.

In several cases there were no local patches of anemia but the left hand looked smaller as well as paler than the right. In fact one marble worker showed a very striking difference in the size of the two hands. Another marble worked showed the reverse condition, a paler and smaller right hand, but this man proved to be left-handed. I also heard from two former granite cutters, not employers, a history of wasting of the muscles of the hand from the use of the air hammer. One of these men used the air hammer for 20 years and after the first 2 years the fingers on the left hand began to get white but later on he noticed that the interosseous muscles (between the bones) of this hand were wasting and deep hollows appeared between the thumb and index finger and between that finger and the next. He gave up the work and the hand gradually went back to its normal state. The second man's history was similar, only that he had the wasting in his right hand, being left-handed and holding the tool in his right hand. He, too, recovered entirely after leaving the work in the shed.

I tried to discover whether this condition persists even after a man has left the trade of stonecutting, whether in cold weather he still has attacks of blanched, numb fingers. There are a few men still working at the trade who say that it passes away gradually; that they used to be much more troubled by it than they are now. On the other hand, I found several men who are no longer cutting stone, who gave up the trade several years ago, and yet still have numb, white fingers at times in winter or if they try to work with cold metal. I will give the essential points in the histories of eight such men.

The first was a granite cutter for 20 years, but left the trade eight years ago. Three fingers of his left hand still get blanched and numb in winter. The second, also a granite worker, stopped work eight years ago but still has the condition at times in the fingers of the right hand (he is left-handed). The third left granite cutting 12 years ago, after four fingers of the left hand had become affected, and those fingers grow white now in winter. The fourth, also a granite worker, came into the office of the union while I was there, and I noticed at once the dead-white shrunken condition of his left hand very characteristic and marked. He told me that he was no longer cutting granite; had not done it for four years, yet the condition persists. Two limestone cutters left the trade six and eight years ago, but both have numb, white fingers on cold mornings. Two marble workers are now cutting sandstone after some years in the marble trade, during which time they suffered from white fingers, and though it is now 10 years and more since they ceased to use the air hammer, they still have traces of the trouble.

It is very important to know whether this condition of the hands affects the men's skill or strength, whether it lessens their earning capacity in case they wish to take up other work than stonecutting. I can not speak with any positiveness as to this, but the tests made by Dr. Rothstein throw some light on the question. Few of the men whom I saw complained of loss of sensation in the fingers great enough to hamper them, except when the fingers were actually numb. The majority noticed no change at all in the intervals between attacks of numbness. But, of course, an occupation which had to be carried on in the cold might be impossible just because the numbness would inevitably come on. For instance, one man had tried to work in an automobile shop, and found that on a cold day he could not pick up or hold small screws or small machine parts with his left hand. Another had taken up work which involved handling a crowbar sometimes, and when he did this in winter his left hand would grow numb and so clumsy that he could not use it with skill. The only men who complained of clumsiness in their own work from numb fingers were marble workers who require a high degree of skill, and who find that cold weather often makes it impossible for them to do their best work, especially if they are out of doors.

To understand why the left hand is most affected by the air hammer, on has only to experiment with it. I tested several hammers, from the small light one used by the carvers to the heavy 1-inch hammer. At first the vibration in both hands is so severe to an unaccustomed person that it is impossible to distinguish much between them, but presently one finds that it is possible to hold the hammer in the right hand without grasping it tightly, while the tool held in the left hand must be clung to with much more force or the blows of the hammer will drive it from the hand. This tool receives the direct blows of the hammer, delivered, I was told, over 3,000 times a minute. The handle of the hammer is large and easy to hold, while the tool is small and the fingers are usually cramped about it which of course drives the blood out of the fingers of the left hand. It is this hand also that does the cutting, guiding the tool along the surface of the stone. The little and ring fingers are pressed against the tool especially close to do this guiding. It did not take me long to become convinced that the effect of the air hammer is greater on the left hand than on the right. The effort expended is greater, the muscles are more cramped, and the vibrations are move violent.

This text of the air hammer explains also why the right hand is less uniformly affected than the left, some men having no trouble with it. The work done by the right hand admits of more variety than that done of the left. The handle may be held in several different ways and the strength of the grip relaxed sometimes. Some men find it possible to relieve the right hand by winding a pad or slipping a piece of hose around the handle, deadened the vibration and protecting the hand from the cold steel. It is true that the right hand does sometimes get the force of the cold air from the exhaust which issues in a sharp blast, freezing cold, from an opening in the handle of the hammer. Some men get this blast on the right thumb or index finger, because in doing certain kinds of work they like to vary the force of the blows from minute to minute, and they can do this by controlling the exhaust with the thumb or finger. The control is usually exerted through a valve lower down in the pipe, which is really the proper way to manage it. However, even when a man does control the exhaust with his thumb or finger, he does not have as much anemia in the right hand as in the left.

PREVALENCE OF WHITE FINGERS AMONG STONECUTTERS

Spastic anemia of the fingers is found to some extent among all the four classes of stone workers, with the exception of those sandstone cutters who have never worked in any other sort of stone.

The first region I visited was the limestone belt of Indiana,--the towns of Bloomington and Bedford. There I was able to examine 21 men who came to the hotel to see me and 17 who were working in the mills at the time I went through them. In Quincy and Barre I saw 50 granite cutters, all but 3 of them at work in the sheds. I examined 78 marble cutters at their work in Proctor, Long Island City, and Baltimore, and 15 sandstone cutters in a mill near Amherst, Ohio. In all, 181 stonecutters and carvers were examined.

Among sandstone cutters spastic anemia of the fingers is not found. I visited a mill in which 15 men were at work at the time and the only ones who had ever had dead fingers were three former marble cutters. They had used the air hammer in marble work and their fingers had shown the effects, but in working with sandstone they had used it so little that the trouble was passing away. The other 12 either used it not at all or very little. This is the rule in sandstone work and the result is that dead fingers are not found among the men in this branch of the stone trade. I found also three marble workers and one granite worker who had not used the air hammer and who also had had no trouble with their fingers.

In limestone, granite, and marble work, the condition is so common that any inquiry about it meets with instant response. One does not have to stop to explain. In Quincy and in Barre many of the granite workers are Italians who speak little English, but as soon as they understood my question they would hold up in answer one, two, three or four fingers of the left hand, but often would shake the head when I then pointed to the right hand. It had been charged that the agitation about this condition among the Indiana limestone men had been influenced by the fact that at the time there was a controversy between the employers and the union concerning the use of the air hammer, but there was no controversy in the granite or marble shops, and yet the testimony given by the men in these two fields was much the same as that given by the limestone men.

On the whole the condition set up in the men's fingers by the use of the air hammer does not seem to be so pronounced among the majority of marble and granite workers as among the limestone cutters. This is true both as regards severity and as regards the proportion of men affected. Among the 38 limestone cutters and carvers there were only 4 who had never had white fingers. About as small a proportion of granite workers had escaped--only out of 50--but the effect was much slower in onset among these men than among limestone workers, most of them having worked 5 years or more before their fingers began to show the effects. This is probably attributable to the fact that granite workers, as I said above, do not usually work all day long with the air hammer; they use the mallet more than do the limestone men. A much larger proportion of marble workers were free from anemia of the fingers--34 out of 78. There were also few men in this last group who complained of the numbness or pain in their fingers or said it made them clumsy or that the vibrations of the machine made them nervous.

I attribute this difference partly to the smaller tool used by most of the marble workers, partly to the fact that marble shops are better heated than the mills and sheds in which the other kinds of stone are cut, and perhaps also to a a greater skill in handling the tool, the carver apparently holding it more easily and lightly than the cutter. However, there is another difference which must also be taken into account. Limestone and granite workers belong to a strong union, while all the marble shops I visited are nonunion, and though I hardly like to say that that fact kept the marble workers from answering my questions as freely as did the union men, still it must be admitted that a visit made to a nonunion shop under the guidance of the foreman or manager is not the best way to elicit full histories of occupational disease.

The slighter effect of the small hammer used by carvers was illustrated in one marble shop where among 12 men who used it only 1 had had white fingers, while among 18 who used the large hammer 12 had had this trouble.

The men who show no effects from the air hammer usually attribute their immunity to a more skillful use of the tool. They say that they hold the chisel lightly and never cramp their fingers around it, or they wind thick cotton or wool around the left hand to protect it from the cold. However, I have seen blanched fingers in many men who wore thick gloves on the left hand. One marble worker told me that the condition of the machine made a great deal of difference. He always had trouble with his hands if he worked in a shop where they used old hammers and did not keep them in order, for these grow loose and the tool slips unless it is held tightly, and the vibration is worse. With a new small tool he has no trouble at all. Certainly some men can use the air hammer for long periods and show no effect from it. I saw a limestone carver who had worked with it 23 years, a granite cutter who had done so all day long for 18 years, and two marble cutters who had used it quite steadily for 15 years, and none of them had any numbness of the fingers.

CAUSE OF SPASTIC ANEMIA OF THE FINGERS

There can be no reasonable doubt that this spastic anemia of the stonecutters' fingers is caused by the use of the air hammer. The more continuously it is used, the greater the number of men affected by numb fingers and the more pronounced their symptoms. The greatest complaint is heard from the limestone and marble cutters who use the air hammer for the greater part of their working time. Granite workers have numb fingers also but they do not seem to experience as much discomfort as do the men in the other two branches, and granite workers do not use the hammer much more than half their time. Sandstone cutters see it little or not at all and this is the one stone trade in which numb fingers are almost unknown. One man only was found in any of the stone trades who had this condition of the fingers and had not used the air hammer, and he had only the tip of his little finger affected.

It is, of course, possible that the same condition might be brought about by the use of the mallet and chisel, for the factors that operate in air-hammer work are present to some degree in work with the mallet--a cold tool grasped tightly by the left hand and vibrating from the blows of the mallet. Only the condition would certainly come on much more slowly and probably never attain the same degree, because handwork can never be as continuous as machine work, it must be interrupted from minute to minute; indeed it is probable that between each two blows of the mallet the hand holding the chisel relaxes slightly. In work with the air hammer the fingers of the left hand grasp the chisel so tightly and so long that I have seen men obliged to bend them back with the help of the right hand after the tool was laid down they had grown so stiff. If is certainly true that while stonecutters may have had "dead fingers" before the air hammer was introduced, the condition was not at all common or striking.

There seem to be three factors in the production of this vascular spasm--cold, cramping of the fingers, and the vibration of the hammer and tool. As we have seen, cold acts as the exciting cause of an attack, but it alone can not cause the trouble. Frostbite is a totally different condition. The clutch of the fingers around the tool drives the blood out of the blood vessels and anything that makes it necessary to hold the chisel especially tight seems to add to the numbness and deadness of the fingers. The large tool has to be held very tightly and marble workers find it more trying than the small tool. Old, loose hammers give more trouble than new, tight ones that hold the tool better. The difference in its effect on different individuals is also unusually explained by the different way they hold the tool, for if it is held loosely the man may have no trouble at all with his fingers.

It is hard to say how much the vibration has to do with it. Men who suffer through nervousness which they attribute to the use of the air hammer think it is the vibration that has caused it, but as to the effect on the hands, it seems impossible to separate the part played by vibration from the part played by the strong muscular contraction. The tool that vibrates most has to be held tightest, so the two factors act together. Dr. Rothstein regards the vibration as probably the important factor in the causation of the vasomotor and sensory disturbances he found in the men examined.

Because of the fact that the use of tobacco sometimes causes a spastic anemia of a distinctly localized character, Dr. Miller in his letter to me about the patient whom he had seen suggested that this point be inquired into when histories of the men were taken. I could not establish any connection between the use of tobacco, smoking or chewing, and the appearance of dead fingers. The men are not, on the whole, heavy users of tobacco, for smoking at work is generally forbidden and chewing during work is not common. I could not find any indication that tobacco was ever the exciting cause of an attack of white fingers.

German and Austrian stonecutters in the limestone region say that the use of the air hammer in those two countries has been forbidden by law because it was discovered that the men who had used it could not handle their rifles properly when doing their military service. There is nothing in German factory inspection reports to confirm this statement. The last reports accessible to us are for the year 1913, and they show that a pneumatic tool for stonework was introduced several years before this in Germany and its use had increased so much that factory inspectors were instructed to gather information as to its effect on the health of stone workers. No description of the tool used is given, but it is clear that it is a drill for use in quarries, probably a drill for blasting charges and for breaking stone. The machines are worked by steam, compressed air, or electricity.

The inspectors who report on them are concerned with the great increase in stone dust occasioned by their introduction and with the possibility of using water to keep this dust down. Only four times in more than 20 reports is there any mention of the vibration of the drill. In some places the men are said to use it only a few hours in the day or to alternate with other work. This last precaution is insisted on by the inspector for Wiesbaden in his district. It is very evident that the tool in use is not an air hammer for working up stone. In the report from one district only, Bremen, there is a passing mention of the occasional use of pneumatic machine in a marble shop.

SUMMARY

Among men who use the air hammer for cutting stone there appears very commonly a disturbance in the circulation of the hands which consists in spasmodic contraction of the blood vessels of certain fingers, making them blanched, shrunken, and numb.

These attacks come on under the influence of cold, and are most marked, not while the man is at work with the hammer, but usually early in the morning or after work. The fingers affected are in right-handed men the little, ring, middle, and more rarely the index of the left hand, and the tips of the fingers of the right hand with sometimes the whole of the index finger and sometime the thumb. In left-handed men this condition in the two hands is reversed.

The fingers affected are numb and clumsy while the vascular spasm persists. As it passes over there may be decided discomfort and even pain, but the hands soon become normal in appearance and as a usual thing the men do not complain of discomfort between the attacks. There seems to be no serious secondary effects following thee attacks.

The condition is undoubtedly caused by the use of the air hammer; it is most marked in those branches of stonework where the air hammer is most continuously used and it is absent only where the air hammer is used little or not at all. Stonecutters who do not use the air hammer do not have this condition of the fingers.

Apparently once the spastic anemia has been set up it is very slow in disappearing. Men who have given up the use of the air hammer for many years still may have their fingers turn white and numb in cold weather.

According to the opinion of the majority of stonecutters the condition does not impair the skill in the fingers for ordinary interior stone cutting and carving, but may make it impossible for a man to do outside cutting in cold weather or to take up a skilled trade which exposes the hands to cold.

The trouble seems to be caused by three factors--long-continued muscular contraction of the fingers in holding the tool, the vibrations of the tool, and cold. It is increased by too continuous use of the air hammer, by grasping the tool too tightly, by using a worn, loose air hammer, and by cold in the working place. If these features can be eliminated the trouble can probably be decidedly lessened.

Subject Index

O

P

Q

R

S

T

V

W

X